Nanotechnology in Undergraduate Education

ACS SYMPOSIUM SERIES **1010**

Nanotechnology in Undergraduate Education

Kimberly A. O. Pacheco, Editor
University of Northern Colorado

Richard W. Schwenz, Editor
University of Northern Colorado

Wayne E. Jones, Jr., Editor
Binghamton University (SUNY)

Sponsored by the
ACS Division of Chemical Education

American Chemical Society, Washington DC

Library of Congress Cataloging-in-Publication Data

Nanotechnology in undergraduate education / [edited by] Kimberly Pacheco, Richard Schwenz, Wayne E. Jones, Jr. ; sponsored by the ACS Division of Chemical Education.
    p. cm. -- (ACS symposium series ; 1010)
    Includes bibliographical references and indexes.
    ISBN 978-0-8412-6968-2 (alk. paper)
    1. Chemistry--Study and teaching (Graduate)--United States--Congresses. 2. Nanotechnology--Study and teaching (Graduate)--United States--Congresses. 3. Nanotechnology--Experiments--Congresses. I. Pacheco, Kimberly. II. Schwenz, Richard W. (Richard William), 1955- III. Jones, Wayne E. IV. American Chemical Society. Division of Chemical Education.
    QD47.N36 2009
    620'.50711--dc22
                              2009024956

The paper used in this publication meets the minimum requirements of American National Standard for Information Sciences—Permanence of Paper for Printed Library Materials, ANSI Z39.48–1984.

Copyright © 2009 American Chemical Society

Distributed by Oxford University Press

All Rights Reserved. Reprographic copying beyond that permitted by Sections 107 or 108 of the U.S. Copyright Act is allowed for internal use only, provided that a per-chapter fee of $40.25 plus $0.75 per page is paid to the Copyright Clearance Center, Inc., 222 Rosewood Drive, Danvers, MA 01923, USA. Republication or reproduction for sale of pages in this book is permitted only under license from ACS. Direct these and other permission requests to ACS Copyright Office, Publications Division, 1155 16th Street, N.W., Washington, DC 20036.

The citation of trade names and/or names of manufacturers in this publication is not to be construed as an endorsement or as approval by ACS of the commercial products or services referenced herein; nor should the mere reference herein to any drawing, specification, chemical process, or other data be regarded as a license or as a conveyance of any right or permission to the holder, reader, or any other person or corporation, to manufacture, reproduce, use, or sell any patented invention or copyrighted work that may in any way be related thereto. Registered names, trademarks, etc., used in this publication, even without specific indication thereof, are not to be considered unprotected by law.

PRINTED IN THE UNITED STATES OF AMERICA

About the Cover: *High Resolution TEM of a 20 nm gold nanoparticle supported on a 150 nm diameter titania (anatase) fiber showing the 0.23 nm interatomic distance of the Au (111) atoms. Image was taken by Mr. Dickson Andala at Binghamton University (SUNY).*

Foreword

The ACS Symposium Series was first published in 1974 to provide a mechanism for publishing symposia quickly in book form. The purpose of the series is to publish timely, comprehensive books developed from the ACS sponsored symposia based on current scientific research. Occasionally, books are developed from symposia sponsored by other organizations when the topic is of keen interest to the chemistry audience.

Before agreeing to publish a book, the proposed table of contents is reviewed for appropriate and comprehensive coverage and for interest to the audience. Some papers may be excluded to better focus the book; others may be added to provide comprehensiveness. When appropriate, overview or introductory chapters are added. Drafts of chapters are peer-reviewed prior to final acceptance or rejection, and manuscripts are prepared in camera-ready format.

As a rule, only original rescarch papers and original review papers are included in the volumes. Verbatim reproductions of previous published papers are not accepted.

ACS Books Department

Contents

1. Overview. 1
   Pacheco, Schwenz, and Jones

Course and Curriculum Innovations

2. Nanotechnology items for ACS Exams standardized tests 7
   T. Holme

3. Seeding nanoscience throughout the undergraduate chemistry curriculum 19
   B. H. Augustine

4. Integration of nanoscale science and technology into undergraduate curricula 49
   S. Iyer

5. Big emphasis on a small topic: Introducing nanoscience to undergraduate science majors 65
   R. Schwenz

Nanoscale Laboratory Experiences

6. X-ray Diffraction 75
   R. Macaluso

7. Development of hands-on nanotechnology content materials: Undergraduate chemistry and beyond 87
   S. C. Larsen

8. Laboratory modules on environmental impacts of nanotechnology 101
   J. Zhang

9. Scanning tunneling microscopy and single molecule conductance: Novel undergraduate laboratory experiments 123
   E. V. Iski

10. **Evolution of a nanotechnology lab for 1st year college students**......... 135
    K. Winkelmann

11. **Integration of nanoscience into the undergraduate curriculum via simple experiments based on electrospinning of polymer nanofibers** 155
    N. J. Pinto

12. **Exploring the Scanning Probe: A simple hands-on experment simulating the operation and characteristics of the atomic force microscope** 167
    A. Layson (Teeters)

13. **Benchtop nanoscale patterning experiments** 177
    T. Odom

Indexes

Author Index 191

Subject Index 193

Chapter 1

Overview

Kimberly A. O. Pacheco[1*], Richard W. Schwenz[1], and Wayne E. Jones, Jr.[2]

1. School of Chemistry and Biochemistry, University of Northern Colorado, Greeley, CO 80639
2. Department of Chemistry and Institute for Materials Research, Binghamton University (SUNY), Binghamton, NY 13902
* Corresponding author

    Nanotechnology is the study, measurement, and manipulation of structures between 1 and 100 nm in size for the development of materials, devices, and systems with fundamentally new properties and functions. (*1, 2*) Since the implementation of the National Nanotechnology Initiatiative in 2001, nanotechnology has been one of the fastest growing areas of international research and development. (*3*) Nearly 10 billion dollars have been allotted to the National Nanotechnology Initiative since its inception in 2001, and it is estimated that upwards of $1.4 billion dollars will be spent on research and education efforts in nanotechnology in 2008 alone, increasing to over $1.5 billion in 2009. (*4*) By 2015, it is projected that more than 7 million jobs in the global market could be affected by nanotechnology. (*2*)

    One of the four stated goals of the National Nanotechnology Initiative is to develop and sustain educational resources, a skilled workforce, and the supporting infrastructure and tools to advance nanotechnology. Some questions currently concerning educators interested in developing and maintaining an educated workforce include 1) where will these workers come from, 2) what backgrounds will be required for them to succeed in the workforce, and 3) do they need a breadth or depth of knowledge regarding nanotechnology or a specific subfield in the discipline?

    Education in the area of nanotechnology has been a rapidly growing area, as well. (*5*) Granting agencies have funded efforts to develop and distribute educational materials to K-20 through formal and informal avenues.

Workshops, symposia, conferences, and panels have been held to discuss best practices for nanotechnology education. Slowly, texts are beginning to surface that address teaching these concepts at the undergraduate level. (*6, 7*) All of these efforts are part of the attempt to increase the emphasis on nanoscience concepts in our educational systems.

In approaching these questions, the general consensus has been that more emphasis needs to be placed on the fundamental concepts of nanoscience earlier in the educational experience. Students need to achieve an understanding of the basic ideas of matter and the structure of matter before an understanding of the micro and macrostructures can be attained.

One of the most difficult issues to tackle with students is the concept of size, particularly when it involves 10^{-9} m. This problem has been well documented in the literature. (*8-10*) As a result, there are many ideas as to the most effective method to tackle this problem. Undergraduate educators approach the problem differently, depending upon if their audience is predominately STEM majors or non STEM majors. Educators interested in each level of education also approach the problem differently. In addition, the modes and methodologies utilized depend upon the audience's interests.

This text grew out of a symposium on nanotechnology held at the American Chemical Society national meeting in August of 2007 and offers a modest attempt at providing ideas and best practices for incorporation of nanotechnology into the undergraduate curriculum. Many of the development programs included herein were funded by the National Science Foundation. The book is divided into two main sections, I. *Course and Curriculum Innovations* and II. *Nanoscale Laboratory Experiences*. The main focus of the text is to facilitate incorporation of nanoscience concepts into the undergraduate curriculum at every level. In many instances, these concepts are already being taught in the classroom, but the scale aspect is not emphasized or the unique properties of specific materials at the nanoscale are not discussed. The ensuing chapters present materials and methods in place at different undergraduate institutions addressing nanoscience.

Section I, *Course and Curriculum Innovations*, takes a broader approach and surveys curriculum modifications implemented at various institutions. This section provides information regarding testing students on nanoscale science concepts, ideas on how to incorporate nanoscience concepts across the curriculum, and some overall effects seen at institutions leading the way for this reform.

Section II, *Nanoscale Laboratory Experiences*, contains detailed experiments appropriate for incorporation into an undergraduate laboratory. Included are a variety of experiment levels, specifically those targeting freshmen to seniors and nonscience majors to STEM majors. These chapters are not intended to provide background for research in the area of nanotechnology, but target the audience most interested in incorporating the laboratories into an undergraduate setting.

Collectively, these chapters provide a foundation for the interweaving of current methods and newly developed technological ideas into the undergraduate curriculum. The scope of these chapters is by no means all encompassing or highly detailed in the scientific basis presented. They are meant to be useful additions in an undergraduate setting and may be modified to fit the individual instructor's needs.

References

1. *Nanostructure Science and Technology*; Siegel, R.W., Hu, E., Roco, M.C., Eds.; Springer: Dordrecht, Netherlands, 1999 (also available at http://www.wtec.org/loyola/nano/).
2. Roco, M.C., The National Nanotechnology Initiative: Past, Present, Future. In *Handbook on Nanoscience, Engineering and Technology*; Goddard, W.A., III; Brenner, D.W.; Lyshevski, S.E.; Iafrate, G.J., Eds.; 2nd ed.; CRC Press: Boca Raton, FL, 2007.
3. Roco, M.C. *Intl. J. Eng. Educ.* **2002**, *18*, 488-497.
4. National Nanotechnology Initiative web site. http://www.nano.gov (accessed April, 2009).
5. NanoEd Resource Portal. http://www.nanoed.org/ (accessed April, 2009).
6. Ratner, M.A., Ratner, D. *Nanotechnology: A Gentle Introduction to the Next Big Idea*; Prentice Hall: Upper Saddle River, NJ, 2003.
7. *Nanoscale Science and Engineering Education*; Sweeney, A.E., Seal, S., Eds.; American Scientific: Stevenson Ranch, CA, 2008.
8. Zhong, C.-J., Han, L., Maye, M. M., Luo, J., Kariuki, N. N., Jones, W. E. *J. Chem. Educ.* **2003**, *80*, 194-197.
9. Tretter, T.R., Jones, M.G., Andre, T., Negishi, A., and Minogue, J. *J. Res. Sci. Teach.* **2006** *43*, 282-319.
10. Jones, M.G., Taylor, A., Minogue, J., Broadwell, B., Wiebe, E., and Carter, G. *J. Sci. Educ. Tech.* **2007** *16* 191-202.

Course and Curriculum Innovations

Chapter 2

The Impact of Nanoscience Context on Multiple Choice Chemistry Items

[1]Karen Knaus, Kristen Murphy and [2]Thomas Holme*

Department of Chemistry and Biochemistry, University of Wisconsin – Milwaukee, Milwaukee, WI 53211

Modern chemistry topics are often introduced into classrooms long before they appear in standardized exams. This paper investigates the role of a nanoscience context in multiple choice items by using a comparative description of the cognitive load effects of such items on a practice exam that was given to students in 2nd-semester general chemistry and 1-semester pre-engineering general chemistry. It includes a classroom comparison study of performance and mental effort analyses in twelve chemistry content areas including a nanoscience and materials context category. In addition, cognitive load effects of paired items in four subcategories were evaluated. Results from the study shed light on the cognitive load effects of nanoscience and materials exam items when these contexts are included within the undergraduate general chemistry classroom.

1. Present address: Department of Chemistry, Univerisity of Colorado – Denver, Denver, CO
2. Present address: Department of Chemistry, Iowa State University, Ames, IA

Introduction

The emphasis for including novel arenas of technological advancement including nanotechnology and green chemistry within the chemistry classroom has been fueled primarily by society's call for more future innovators to propose global solutions to environmental, medical and engineering issues that have long continued to peril society. To encourage research and development in the field of nanotechnology, the U.S. government established the National Nanotechnology Initiative (NNI) in 2001, whose annual budget has nearly tripled since its inception (1).

The option of including nanoscience within an individual course essentially lies within the purview of the instructor, but there are hurdles to that inclusion on a large scale, including such things as nationally normed exams. In chemistry, such exams have been produced for over 70 years by the Examinations Institute of the Division of Chemical Education of the American Chemical Society. National exams invariably reflect a conservative assessment of content coverage, so new material, such as nanoscience, is often slow to appear. In principle, it is easier to include new content by providing context for "traditional" coverage rather than testing content that is specific to the emerging field – in this case nanoscience. In order to carry out such context based inclusion, however, it is important to understand the impact of the context, and the study reported here investigates this question utilizing a novel analysis based on cognitive load theory. To date, no other published work exists in the literature which involves the use of nanoscience and special materials assessment items to examine the cognitive implications of various chemistry classrooms with respect to their inclusion or omission of these learning contexts.

Cognitive load theory may be described as the amount of mental activity imposed on the working memory at any instance in time. This concept is arguably descendent from the seminal paper by Miller in 1956 (2) which proposed that the human cognitive system can actively process 7 ± 2 pieces of information at any time. While direct measures of cognitive load are a challenge, the studies of Gopher and Braune (1984) indicated that "subjects can subjectively evaluate their cognitive processes, and have no difficulty in assigning numerical values to the imposed mental workload or the invested mental effort in a task (3)" (4, p. 739). Although mental effort has been measured using various techniques (5, 6), including; rating scales, physiological techniques (i.e. heart-rate variability and pupil dilation response (7), and dual-task methods), subjective measures have been found to be very reliable, unobtrusive and very sensitive (8-11). "The intensity of effort expended by students is often considered the essence of cognitive load (12, 8)" (10, p. 420), and most studies for measuring mental effort have used a subjective rating system (13).

Combined measures of performance and mental effort can be used as a tool to help us learn more about instructional efficiency (4). In our previous work (14), we established a method for determining cognitive efficiency in different chemistry categories using a combination of performance and mental effort measures collected from three large general chemistry classrooms who took a practice exam. In this paper, we present the results of a synonomous research study that was carried out in conjunction with our cognitive efficiency analysis, where we investigated the cognitive load impacts of nanoscience context on multiple-choice exam items for "standard" general chemistry and pre-engineering general chemistry students. Results from this study provide new insight into the cognitive load consequences of nanoscience context on test items when students are introduced to such topics in their chemistry classrooms.

Instrument Design

The practice exam instrument used in our study utilizes 50 multiple choice items, including 6 items specifically keyed to materials science and/or nanoscience. After each exam item, a mental effort item was inserted into the exam format that asked students to introspect on the degree of mental effort expended on the previous question answered (Figure 1). We used a 5-point Likert scale consistent with the number of available multiple choice options found on a typical scantron answer key.

How much mental effort did you expend on question #1
Very little
Little
Moderate amounts
Large amounts
Very large amounts

Figure 1. Example of mental effort item inserted into the practice exam format.

Two nanoscience items in the sub-categories of spectroscopy and band theory, as well as two materials items in the sub-categories of intermolecular forces (IMF) and Lewis structures, were inserted into the practice exam to serve as paired items to non-nanoscience/materials items on the exam in the same sub-categories.

Exam Administration

The practice exam was given approximately one week prior to final examinations in courses at an urban university in the Midwest. Data described here arises from performances from a total of 158 students who took the practice exam (83 students in 2nd-semester general chemistry and 75 students in the single-semester pre-engineering chemistry course) and further agreed to participate in the research component of the study by signing the relevant IRB consent form. Students received individualized email feedback within one day of taking the practice, which included both their performance and mental effort averages in twelve chemistry content areas including the nanoscience and materials content area. Aside from research purposes, the information provided from the practice exam was primarily meant to help students validate their current levels of content knowledge, cognitive resource usage, and potentially guide them in developing a better study plan to prepare for their upcoming final exam the following week.

Data Analysis

Cognitive efficiency analysis, as described in our previous work (14) allowed us to probe differential performance and mental effort effects for all items on the practice exam, including the nanoscience and materials science items for the two different types of general chemistry courses (Figures 2 & 3).

When comparing the cognitive efficiency graphs (Figures 2 & 3) for the two different types of chemistry courses (2nd-semester general chemistry vs. 1-semester pre-engineering survey course), individual differences in terms of the relationship between performance and utilization of mental resources in different chemistry areas can more clearly be identified. Both courses (2nd-semester general chemistry and single-semester pre-engineering chemistry) show low cognitive efficiency in the solutions category, an observation that may be important for instructors because it is apparently robust regardless of the course emphasis or instructor. Given the different emphasis of the course and time for instruction, comparisons between the one-semester course for engineering students and the traditional two-semester general chemistry sequence courses provides additional observations. First, the high cognitive efficiency in the nanoscience and materials chemistry area for the single-semester pre-engineering chemistry course is achieved despite the "survey"

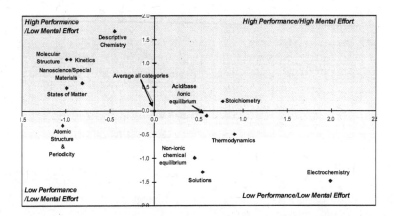

Figure 2. Graph of cognitive efficiency in different chemistry categories for 1-semester pre-engineering chemistry course. Normalized values (a.k.a. standardized or "z scores") for classroom performance in each chemistry category are plotted against normalized values for average mental effort in each category. Values in the upper left quadrant indicate content areas of high performance and low mental effort, values in the upper right quadrant indicate content areas of high performance and high mental effort, values in the bottom left quadrant indicate areas of low performance and low mental effort, and lastly, values in the lower right quadrant of the graph indicate areas of low performance and high mental effort.

nature of this course. This observation supports the establishment of a hypothesis that introduction of modern concepts into general chemistry such as nanoscience and materials may prove effective even if the time invested is modest (no topics receive large allocations of time in the one-semester survey course).

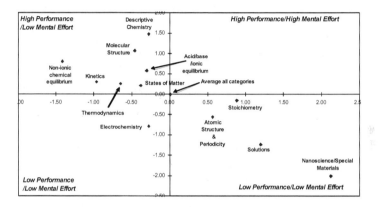

Figure 3. Graph of cognitive efficiency in different chemistry categories for 2nd-semester general chemistry students (average of two 2nd-semester general chemistry course sections).

Paired Item Analysis

The second component of analysis for determining the cognitive load effects of nanoscience and materials science contexts on exam items for the two different chemistry courses, involved a paired item analysis. Two nanoscience items in the sub-categories of spectroscopy and band theory, as well as two materials items in the sub-categories of intermolecular forces (IMF) and Lewis structures, were inserted into the practice exam to serve as paired items in these same sub-categories. The first comparison we analyzed was both groups of students' differential performance on non-nanoscience and materials exam items versus nanoscience and materials items (Figures 4 & 5).

In Figure 4, the performance comparison on non-nanoscience and materials & nanoscience and materials items for pre-engineering chemistry students, it can be seen that in both the areas of spectroscopy and Lewis structure, the engineering students have a higher performance when the format of the item includes reference to either nanoscience or special materials. This observation is very interesting, because unlike the areas of intermolecular forces and band theory, neither of these concepts were known to be taught in the context of

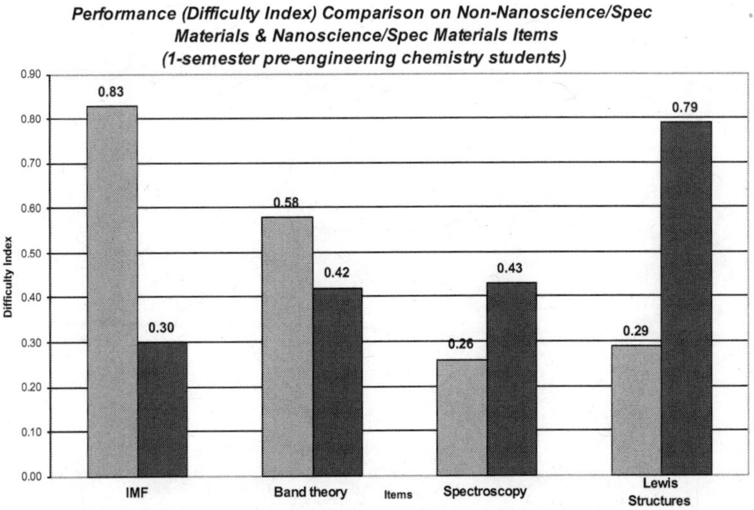

Figure 4. Performance comparison on non-nanoscience and materials (grey) and nanoscience and materials (black) exam items (1-semester pre-engineering chemistry students).

nanoscience or materials (either during the lecture component or textbook contents).

In Figure 5, the performance comparison on paired items for 2nd-semester general chemistry students, it can be seen that in all four areas the students perform better when the items are not presented within the context of nanoscience or materials.

The discrepancies found between performances of 2nd-semester general chemistry students and pre-engineering chemistry students on contextualized exam items suggest something about the role various learning environments will play on the performance of students on these particular items. Although content coverage in the two general chemistry courses was similar, four things were notably different, (1) different instructors taught the courses; (2) the single-semester pre-engineering course was a survey course with engineering applications; (3) it did not cover the traditional chemistry content in as much depth as two semesters of general chemistry; and (4) the textbook used for the pre-engineering general chemistry course included nanoscience and materials context. Regardless of these differences between the two courses, the pre-engineering general chemistry students were not introduced to the concepts of spectroscopy and Lewis structures within the context of either the nanoscience or materials. Furthermore, these results hint at the possibility that a greater transfer of knowledge in these chemistry areas was achieved due to the nanoscience and materials contextual emphasis inherent in the pre-engineering chemistry course in comparison to the regular general chemistry course.

The practice test instrument also allows a comparison of student perceived mental effort as shown in Figures 6 and 7.

The average mental effort comparisons in Figures 6 and 7, indicate differential cognitive load effects experienced by the two different groups of students (1-semester pre-engineering chemistry students versus 2nd-semester general chemistry students) on paired items.

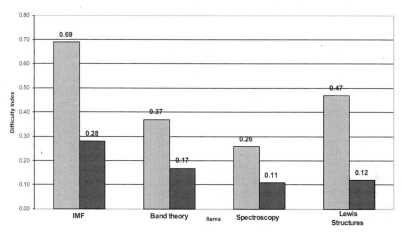

Figure 5. Performance comparison on non-nanoscience and materials (grey) and nanoscience and materials (black) exam items (2nd-semester general chemistry students).

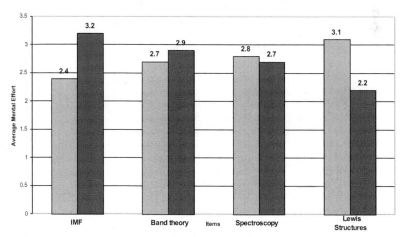

Figure 6. Average mental effort comparison on non-nanoscience and materials and nanoscience and materials exam items (1-semester pre-engineering chemistry students).

For the area of intermolecular forces, both groups of students perceived the nanoscience materials item to be more mentally challenging. For the area of band theory, the 2nd-semester general chemistry students perceived the nanoscience and materials item to be substantially more mentally challenging than the pre-engineering general chemistry students. This result is not surprising as band theory is a topic which is more heavily emphasized in the pre-engineering general chemistry course than in other general chemistry courses. For the area of spectroscopy, the 2nd-semester general chemistry students indicated that they exerted less mental effort when attending to the nanoscience and materials item (average mental effort of 2.4 item) than the paired item (average mental effort of 3.0). The most interesting aspects of the results from the spectroscopy paired items is that although the pre-engineering chemistry students experienced a negligible difference in cognitive load for the paired spectroscopy items, they performed substantially better than the 2nd-semester non-engineering general chemistry students on the nanoscience and materials item. Again, the chemistry concept of spectroscopy was not taught in either the context of nanoscience or materials in the pre-engineering course. This result bolsters the suggestion that the pre-engineering chemistry students may have experienced improved transfer test performance gain for the topic of spectroscopy over the 2nd-semester general chemistry students due to the general inclusion of this context elsewhere in the course.

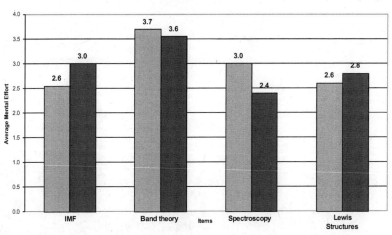

Figure 7. *Average mental effort comparison on non-nanoscience and materials and nanoscience and materials exam items (2nd-semester general chemistry students).*

In the last area, Lewis structures, the average mental effort data allows us to explore the differential cognitive load effects of the Lewis structure paired items. The pre-engineering students appear to be exerting less mental effort for the nanoscience and materials Lewis structure item relative to the paired item.

For the 2nd-semester general chemistry student, the cognitive load effects of the Lewis structure paired items are quite comparable.

Conclusion

This paper investigates the performance and cognitive load effects of nanoscience and materials exam items for 1-semester pre-engineering students and 2nd-semester general chemistry students. The performance and mental effort histograms and cognitive efficiency analyses confirmed that students in the 1-semester pre-engineering chemistry course who are exposed to nanoscience and materials context in both their classroom and textbook material, in general have both higher performance and greater cognitive efficiency in this area.

Performance and cognitive load effects of paired items in four subcategories (intermolecular forces, band theory, spectroscopy, and Lewis structures) were evaluated and compared for 1-semester pre-engineering students and 2nd-semester general chemistry students. Evidence from these studies suggest that for a course where students have some contextualization utilizing nanoscience, no differences arise when the exam items contained nanoscience and materials context. In contrast, for a course that does not include any nanoscience contextualization students performed much better on the non-nanoscience items than nanoscience and materials items. The average mental effort (a.k.a cognitive load data) from the paired items indicated that nanoscience and materials items imposed the same amount of cognitive load as the non-nanoscience and materials item for the pre-engineering students. For the 2nd-semester general chemistry students, on average, the cognitive load differences of nanoscience and materials items and non-nanoscience and materials items (for these 8 items on the practice exam) were minimal. Furthermore, the nanoscience and materials items that were included on the practice exam used in our study, did not impose differential cognitive load effects on one population of students over another. These results suggest that although including nanoscience and materials items on multiple choice exams may put students who have been exposed to the nanoscience context at some level in the classroom (either through classroom discussions or integration in the textbook) at a slight advantage, these items on average do not impose additional cognitive load for regular general chemistry over students in a 1-semester pre-engineering chemistry survey course.

These results, for a limited set of nanoscience context items, suggest that the addition of nanoscience to national exams can be accomplished. It is, however, likely important that instructors look to include nanoscience context in their courses. Even when the inclusion of nanoscience is not directly related to the topic tested, exposure to the idea appears to help students in terms of both exposure and mental effort.

Suggestions to Instructors

There are a number of methods a chemistry instructor can use to begin to incorporate non-traditional context such as nanoscience, special materials and green chemistry content into their classrooms. Such methods include incorporation of textbook and/or laboratory manual material, on-line resource information (i.e., of which several excellent examples exist) and may include such things as general information, animations, teaching modules, laboratory modules, use of remotely accessible instrumentation for studying such materials, and even outreach activities. Of course, students are always welcome and can be invited by chemical educators to further entertain their own curiosities of such topics through use of these resources as well. A list of some of the available resources for inclusion of nanoscience and materials context in the chemistry classroom can be found below.

Books (some of which may have accompanying lab manuals):

Chemistry for Engineering Students by Lawrence Brown (Author) and Thomas Holme (Author)

Tomorrow's Chemistry Today: Concepts in Nanoscience, Organic Materials and Environmental Chemistry by Bruno Pignataro (Editor)

Single Molecule Chemistry and Physics: An Introduction (NanoScience and Technology) by C. Wang (Author) and C. Bai (Author)

Introduction to Materials Chemistry by Henry R. Allcock (Author)

Functional Molecular Nanostructures (Topics in Current Chemistry) by A Dieter Schliter (Editor)

Introduction to Nanoscience by Gabor L Hornyak (Author), H.F. Tibbals (Author), J. Dutta (Author), and A. Rao (Author)

Nanochemistry by G.B. Sergeev (Author)

NanoBioTechnology: BioInspired Devices and Materials of the Future by Oded Shoseyov (Editor) and Ilan Levy (Editor)

Nanoscale Materials in Chemistry by Kenneth J. Klabunde (Editor)

On-line resources:

http://invsee.asu.edu/Invsee/invsee.htm
IN-VSEE Interactive Nanovisualization in Science & Engineering Education web-site

http://mrsec.wisc.edu/Edetc/
Materials Research Science and Engineering Center on Nanostructured Interfaces

http://www.nanoed.org/index.shtml
NanoEd Resource Portal

http://www.cns.cornell.edu/index.html
Center for Nanoscale Systems

http://www.nnin.org/nnin_edu.html
National Nanotechnology Infrastructure Network Education Portal

Acknowledgments

This work is based on work supported by the National Science Foundation (NSF) under grant numbers CHE-0407378 and DUE-0618600. Any opinions, findings, and conclusions communicated in this material are those of the authors and do not necessarily reflect the views of the NSF. The authors would also like to thank instructors at UWM who allowed for the collection of data in their classrooms, which contributed to our research findings.

References

1. National Nanotechnology Initiative Funding Page. http://www.nano.gov/html/about/funding.html (assessed July 2008).
2. Miller, G.A. *Psychol Rev.* **1956**, *101*, 343-352.
3. Gopher, D.; Braune, R. *Hum Factors.* **1984**, *26*, 519-532.
4. Paas, F. G. W. C.; Van Merriënboer, J. J. G. *Hum Factors.* **1993**, *35*, 737-743.
5. *Mental Workload: Its theory and application;* Moray, N., Ed.;. Plenum Press: New York, 1979.
6. Eggemeier, F. T. Properties of workload assessment techniques. In *Human mental workload;* Hancock, P. A., Meshkati, N., Eds.; Elsevier: Amsterdam, The Netherlands, 1988; pp 41-62.
7. Ahern, S. K.; Beatty, J. *Science.* **1979**, *205, 1289-1292.*

8. Paas, F. G. W. C. *J. Educ. Psychol.* **1992,** *84*, 429-434.
9. Paas, F. G. W. C.; Van Merriënboer, J. J. G. *J. Educ. Psychol.* **1994,** *86*, 122-123.
10. Paas, F. G. W. C.; Van Merriënboer, J. J. G.; Adam, J. J. *Percept Mot Skills,* **1994,** *79,* 419-430.
11. Ayres, P. *Learn Instr.* **2006,** *16,* 389-400.
12. Hamilton, P. Process entropy and cognitive control: Mental load in internalized thought process. In *Mental workload: Its theory and measurement;* Moray, N., Ed.; Plenum Press: New York, 1979; pp 289-298.
13. Paas, F. G. W. C.; Renkl, A.; Sweller, J. *Educ Psychol.* **2003,** *38,*1-4.
14. Knaus, K. J.; Murphy, K. L.; Holme, T. A. *J. Chem. Educ. submitted.*

Chapter 3

An Evolutionary Approach to Nanoscience in the Undergraduate Chemistry Curriculum at James Madison University

Brian H. Augustine, Kevin L. Caran, Barbara A. Reisner

Department of Chemistry and Biochemistry, MSC 4501
James Madison University
Harrisonburg, VA 22807

We describe a model used for seeding nanoscience topics throughout the undergraduate chemistry curriculum at James Madison University (JMU). An overview of the evolutionary changes to the chemistry curriculum as a result of this program will be presented. Lecture topics in general, inorganic, organic, materials science and physical chemistry have been added or improved and laboratories for general, organic and physical, have been developed and implemented. A new general physical science course for nonscience majors and an upper-division majors lecture-laboratory course called, "*Science of the Small: An Introduction to the Nanoworld*" have been developed and will be broadly described. Nanoscience topics from the current scientific literature have been introduced into materials science, physical chemistry lab, and literature and seminar courses. We will further describe how a new series of vertically integrated laboratory experiments exploring the properties of supramolecular micelles are being used to introduce nanoscience into several different courses. In addition, we will discuss how this project, in particular the *Science of the Small* course, has helped to catalyze "top-down" thinking about seeding a range of interdisciplinary topics throughout the undergraduate chemistry curriculum. Finally, we address how this evolutionary approach can be used by non-experts to begin to seed nanoscience topics into the undergraduate chemistry curriculum.

Why Nanoscience in the Undergraduate Chemistry Curriculum

The development of nanoscience and nanotechnology has received considerable attention from both the scientific (*1-3*) and popular media (*4-7*) during the last decade. Proponents hail the manipulation of materials at the nanometer-scale as one of the most significant technological achievements yet devised, while others are more dubious about the potential of self-replicating nanomachines causing an environmental or health crisis (*4, 7*). While both of these extreme scenarios are unlikely, as scientists and engineers continue to unlock the secrets of manipulating matter at the molecular-scale, it becomes increasingly important to educate and prepare a future workforce with an understanding of the basic science and technology underlying the nanotechnology revolution.

Nanoscience by its nature is an interdisciplinary field cutting across traditional areas of chemistry, physics, biology and engineering. In order to develop a supply of undergraduate students prepared to work on nanotechnology problems, the National Science Foundation established the Nanotechnology Undergraduate Education (NUE) program in 2003. Initially the program was designed to fund projects that were one year in length. In 2005, the NUE was expanded to include two year projects when the Department of Chemistry and Biochemistry at James Madison University was funded. The crux of the project was to seed nanoscience topics throughout the undergraduate curriculum in what we described as an *evolutionary* rather than a *revolutionary* approach to nanoscience education. In the remainder of this chapter, we will present an overview of how this program has begun to shape the overall curriculum in chemistry at JMU, interdisciplinary work with other departments at JMU, and preliminary work in the assessment of student knowledge and attitudes about nanoscience.

Why an Evolutionary Approach

Most undergraduate programs in traditional disciplines such as chemistry, physics, and biology focus primarily on building the foundations of these disciplines when designing their curricula. Few departments have made a concerted effort to develop interdisciplinary programs that enable undergraduate students to see the connections between these fields. Nearly 50% of recent doctoral recipients in chemisty received bachelor's degrees at primarily undergraduate institutions (PUIs) (*8*), yet many of these students have had little exposure to coursework or research that cuts across traditional disciplinary boundaries due to a conventional structuring of these programs. An emerging field such as nanoscience and nanotechnology requires that undergraduate science education programs more purposefully and clearly show students how the traditional disciplines intersect. Prior to receiving the NUE grant, efforts at JMU to introduce nanoscience into the undergraduate curricula had been *ad hoc*, with little coordination between faculty or across different courses.

One obvious solution to this problem would have been to design a chemistry program that is heavily nanotechnology-oriented. However, this *revolutionary* approach does not meet the needs or desires of the majority of undergraduate chemistry students. The growing demand for interdisciplinary work in biochemistry, environmental chemistry, forensics, and materials science, in addition to the more traditional subdisciplines in chemistry, made it clear that a radical change in the chemistry curriculum focused solely on nanoscience was unwise. In a large department like the Department of Chemistry and Biochemistry at JMU, there would be little support for focusing on a single area of modern science. A second approach would be to require that all students take a new course in nanoscience. However, this tactic is also problematic because of the large number of existing course requirements in chemistry, its cognate areas, and general education.

We proposed a more *evolutionary* approach that begins by introducing all general chemistry students to nanoscience through laboratory experiments during their first year then reinforcing nanoscience during the second year organic sequence by using nanotechnology-related lecture topics, demonstrations, and laboratory experiments. Molecules synthesized and characterized during the second year organic labs are then discussed or used in nanoscience-themed experiments in subsequent advanced courses, and ultimately in a new upper-division lecture-lab course entitled, *"Science of the Small: An Introduction to the Nanoworld."* This approach was designed so that all students taking chemistry classes would be exposed to nanoscience, and so that nanoscience concepts would build from the first year experience and be reinforced in subsequent courses. In addition, the connection between concepts such as synthetic strategies in organic chemistry and applications and characterization of these molecules in physical chemistry or materials science would become more obvious. Finally, this has resulted in significantly more support from the faculty at large. It should be noted that a similar approach – making connections using nanoscience as a bridge – is now also being employed by Sienna College and the University of Richmond (*9*).

Challenges to Introducting Nanoscience in the Undergraduate Curriculum

While many agree on the importance of exposing our students to nanoscience during their undergraduate years, a number of barriers exist that hinder the introduction of nanoscience into the chemistry curriculum. These challenges are not unique to the field of nanoscience, nor does each apply to every institution. However, we believe that it is important to articulate these challenges. It is easier to develop solutions if the barriers that exist to change are understood and addressed in curriculum development.

The first major challenge to the widespread introduction of nanoscience is the content knowledge of faculty. While Ph.D. granting institutions and many primarily undergraduate institutions (PUIs) have faculty with expertise in nanoscience, a number do not. If a faculty member does not have direct experience with nanoscience, it can be challenging to introduce it into a course because they must operate outside of their "comfort zone." Because faculty tend to be highly specialized experts, they may not have made the connection that the basic principles in their own field are routinely employed in nanoscience. Furthermore, if a single faculty member has expertise in nanoscience, he/she may not know how best to introduce nanoscience into other areas. For example, an organic faculty member with research interests in nanoscience may have difficulty providing a physical chemistry colleague with examples of how to incorporate nanoscience into the physical chemistry laboratory. To surmount this barrier, it is imperative that well-developed classroom and laboratory activities exist so that non-experts can more easily adopt nanoscience activities.

A second challenge, which is particularly acute at larger institutions, is the hurdle that large class sizes presents. At many institutions, including ours, general chemistry programs accommodate hundreds, even thousands of students each year. Ideally, we would like to adopt experiments that have recently been published in the scientific literature. However, there are practical requirements that must be met for large classes which often make the adaptation of recent papers difficult. Chemicals must be easily accessible, inexpensive, have low toxicity, and have minimal disposal costs. Furthermore, there may be restricted access to analytical instrumentation and resources may not be available to handle unstable or sensitive reagents and products. Much of the chemistry described in the literature is highly sensitive to technique, thus leading to problems with reproducibility that will significantly affect student results. It is much easier to develop activities for smaller classes because cost is not as significant an issue and hazards are more easily monitored. Therefore, we recommend the development of greener laboratories that use equipment readily accessible to most chemistry programs which do not employ fastidious systems.

A third challenge that may exist is differences in teaching philosophy. Laboratory experiments can serve a number of purposes: to introduce techniques, to illustrate concepts discussed in lecture, to allow students to construct their own knowledge base through inquiry-based activities, to show applications of chemistry (e.g. nanotechnology, environmental science, forensic science, medicine, etc.), or to increase students' awareness of the role of chemistry in their lives. None of these goals are mutually exclusive, but it is rare that all of these pedagogies can be addressed by a single lab activity. The goals of the instructor and the nature of a laboratory activity must be well-matched for the lab to be viewed as successful by both faculty and students. To address the greatest range of pedagogies, new laboratories should address established core topics and techniques. Most college instructors are adept at rewriting labs to fit their own teaching style.

While we have content experts in nanoscience, we face all of these challenges at JMU. In the next section, we will discuss how we have addressed these issues. We will highlight some of the problems that exist in making these changes at a large institution and discuss the evolutionary changes that we have made to our courses.

Evolutionary Changes to Existing Courses

Fundamentally, an evolutionary approach to nanoscience teaching in the undergraduate chemistry curriculum implies choosing appropriate places where nanoscience topics can be seeded into existing courses with limited new course development. Our belief is that by seeding nanoscience topics into lectures, laboratories, and demonstrations at all levels of the curriculum in self-contained modules that address topics already taught in the standard chemistry curriculum, there would be a smaller barrier to adoption because (1) there would be no need to change the topical material already taught in a course, and (2) faculty with limited expertise in nanoscience would feel more comfortable because they would be teaching discrete modules that build upon standard topics. Success with an evolutionary approach to nanoscience is predicated on the idea that if the awareness of students is raised throughout their undergraduate experience, then at the very least all undergraduate chemistry majors will have received some exposure to nanoscience topics and a greater appreciation of the field. Ideally, these nanoscience experiences early in students' college experience may lead to more students electing to take the *Science of the Small* course. We begin by providing background information about James Madison University and then follow with the courses that have been most affected by our changes.

Background – The Chemistry Department and Major at JMU

In order to better understand the challenges faced when integrating nanoscience throughout the JMU chemistry curriculum, it is helpful to understand the overall structure of the university and the chemistry program. James Madison University is a comprehensive state university and PUI with approximately 18,000 students. By 2013, our projected enrollment is over 21,500 students. Since JMU is a PUI, neither lectures nor laboratories are taught by graduate student assistants. A full-time professional faculty member (non-tenure track) manages the general chemistry laboratory program and coordinates the undergraduate student assistants who prepare solutions and glassware for each of the over thirty weekly laboratory sections. The organic chemistry laboratory is coordinated by an instrumental chemist in consultation with the organic faculty. All other laboratories are coordinated by the faculty teaching the course. Laboratories are taught by tenure-track faculty or experienced lecturers, and undergraduate students serve as in-lab teaching assistants. JMU graduates approximately twenty five undergraduate chemistry majors per year. Thus, with well over a thousand students per year taking general, organic and biochemistry laboratory courses, these represent a significant service component by the department to the university. Without graduate assistants, these laboratory courses represent a major investment in the time of tenure-track faculty.

JMU's current undergraduate chemistry program closely follows the guidelines established by the American Chemical Society's Committee on Professional Training with a required core curriculum consisting of general, organic, biochemistry, inorganic, analytical, and physical chemistry as well as the standard cognates and associated laboratory courses as established by the

ACS prior to 2008. Students are required to take at least one elective course. These electives consist both of advanced courses in core areas, and interdisciplinary courses such as materials science, polymers, lasers, nuclear, and environmental chemistry. The department has a tradition of integrating disciplines in specific courses, as exemplified by our required integrated organic/inorganic laboratory (10), several materials science courses (co-taught between the departments of chemistry and physics), and a biophysical course (co-taught by a biochemist and physicist). However, prior to this project, we had not made a concerted effort to weave a broad interdisciplinary topic throughout the curriculum.

The following sections will provide further details about the courses into which nanoscience concepts have been integrated.

General Chemistry Lecture and Laboratory Courses

With the exception of approximately one dozen well-prepared chemistry majors with passing AP scores, all students who need a full year of General Chemistry enroll in the same two-semester lecture and laboratory sequence. Each three credit-hour lecture (CHEM 131/132) has a co-requisite one credit hour lab (CHEM 131L/132L). All courses are offered every semester. Over the course of an academic year, approximately 850 students take CHEM 131/131L and 600 students take 132/132L. These classes are populated with students from pre-professional and allied health disciplines (pre-medical, pre-dental, pre-veterinary, pre-physical therapy, kinesiology, and dietetics), physical, earth, and life science majors, as well as non-science majors fulfilling part of their general education science requirement. Lecture sections are capped at 150 students, necessitating that several lecture sections be offered in a given semester. Laboratory sections (capped at 24 students) meet for one 2.75 hour period each week. Tenure-track faculty typically teach between one and four sections of general chemistry laboratory each year depending on other disciplinary teaching assignments.

There are several places in the general chemistry curriculum where nanoscience topics logically intersect topics that were taught prior to our efforts to integrate nanoscience into these courses. Table I provides examples of topics where we have integrated nanoscience applications into the teaching of "traditional" general chemistry topics, and the semester that this occurred in our general chemistry sequence.

In addition to examples of nanoscience ideas seeded into the existing curriculum, one of the authors has developed a "special topics lecture" specifically themed around nanoscience. Having the lecture later in the semester of General Chemistry I is important in order for students to grasp the importance of molecular geometry and intermolecular forces in the nanoworld. This lecture highlights recent literature on nanoscience topics, such as the top-down engineering of biologically inspired nanofabricated sensors and carbon nanotubes. A particular emphasis is placed on scaling (what is a nanometer, what is nanotechnology?) as this is a key conceptual difficulty for students (11). A demonstration of atomic force microscopy (AFM) is also performed in this lecture. Figure 1 shows a figure used to illustrate scaling to the general chemistry class. There are other similar examples of scaling available on the internet (12).

Table I. General Chemistry Nanoscience Topics

Topic	Semester	Demonstration, Experiment or Classroom Example
Relative atomic abundance	1st	STM of atomic Cu surface
Balancing equations	1st	CdS nanoparticle synthesis
Atomic / quantum theory	1st	STM, quantum corrall
Light / waves	1st	LEDs vs. incandescent bulbs
Absorption / emission	1st	"Color My Nanoworld" laboratory. Emission from LEDs, glow discharge tubes
Lewis structures, molecular geometry	1st	Amphiphilic molecules, Graphite vs. diamond vs. buckyballs/carbon nanotubes
Polarity of molecules	1st	Amphiphilic molecules, molecular bilayers
Condensed matter	1st	Crystal lattices, Si-diamond lattices
Electrochemistry	2nd	Nanostructured photovoltaics, H_2 fuel cells

Making changes in the general chemistry laboratory is more challenging than making changes in the lecture because of the number of sections involved and the reagent preparation and disposal required. A number of nanoscience activities that can be used in the chemistry laboratory have been developed, but most are geared towards use in a physical, analytical, or instrumental laboratory. A smaller number have been developed for organic, biochemistry, or inorganic chemistry laboratories. Many of these labs could be implemented in a superficial way in the general chemistry lab, but few laboratories have been designed to integrate both nanoscience and the fundamental principles covered in general chemistry. Even fewer can be implemented in large enrollment laboratory courses. A list of laboratory activities that have been developed for use in large general chemistry programs is provided in Table II.

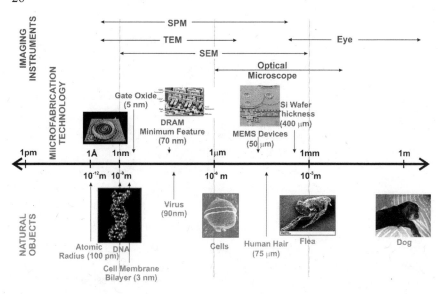

Figure 1. Illustration of scaling used in general chemistry and high school outreach presentations showing natural objects, micro/nanofabrication technologies and imaging technologies. (Images adapted from AIP, IBM Corporation, and Sandia National Laboratory, SUMMiT Technologies, www.mems.sandia.gov)

Table II. Nanoscience Laboratories for General Chemistry for Large General Chemistry Programs		
Laboratory Activity	*General Chemistry Topics*	*Ref.*
Determining the critical micelle concentration of aqueous surfactant solutions: Using a novel colorimetric method	Intermolecular Forces; Organic Chemistry; Properties of Solutions	(13)
Preparation and Properties of an Aqueous Ferrofluid	Crystal Structure; Intermolecular Forces; Magnetism; Stoichiometry; Synthesis	(14)

Chemistry with Refrigerator Magnets: From Modeling of Nanoscale Characterization to Composite Fabrication	Magnetism; Polymers; Synthesis	(*15*)
Replication and Compression of Surface Structures with Polydimethylsiloxane Elastomer	Polymers; Structure – Property Relationships	(*16*)
Color My Nanoworld	Electrolytes / Nonelectrolytes; Intermolecular Forces; Light / Color Relationships; Oxidation / Reduction; Synthesis	(*17*)
Nanopatterning with Lithography	Structure of Solids	(*18*)
Nanoresistors and Single-Walled Carbon Nanotubes: Using An Ohmmeter to Test for Hybridization Shifts	Allotropes; Structure – Property Relationships; Synthesis	(*19*)
Do Nanoparticles Have Equilibrium Constants?	Equilibrium	(*20*)
From Rust to High Tech: Semi-Synthesis of a Ferrofluid Using FeO Nanoparticles	Oxidation / Reduction; Synthesis	(*21*)
Synthesis of Silver Nanoparticles: An Undergraduate Laboratory Using a Green Approach	Spectroscopy; Synthesis	(*22*)
Synthesis and Study of Silver Nanoparticles	Intermolecular Forces; Oxidation / Reduction; Spectroscopy; Stoichiometry; Synthesis	(*23*)
Preparation of CdS Nanoparticles by First-Year Undergraduates	Intermolecular Forces; Light / Color Relationships; Quantum Effects; Spectroscopy; Synthesis	(*24*)
Octanethiol Monolayer on Silver	Intermolecular Forces; Organic Chemistry	(*25*)
Microcontact Printing of Thiols	Intermolecular Forces; Synthesis	(*25*)

We have introduced the published *Color My Nanoworld* experiment (*17*) into the first semester general chemistry course. This experiment involves the synthesis of sodium citrate passivated Au nanoparticles and clearly demonstrates the nanoscience concept that objects on the nanometer scale exhibit different optical properties based on the nanoparticle size. Students prepare a solution of nanoparticles with a size-range causing the solution to appear red. They observe the difference between adding solutions of electrolytes and non-electrolytes. The former shields the surface charge of the nanoparticles, allowing aggregation of the nanoparticles, resulting in a larger apparent diameter, thus causing the solution to change color. Students generally enjoy this experiment, although very few understand the origin of the color change in the Au nanoparticle system. We have not attempted to explain surface plasmon resonance to this audience, but rather focus on the fact that the properties of the nanoparticles (in this case, color) are profoundly affected by their size, and that the nanoparticles can be used as a qualitative probe of the properties of the solution.

Because students found this lab interesting, but saw little connection with the topics being explored in general chemistry, we are working to introduce additional context into this laboratory. We will adopt an implementation to this laboratory being developed by the Rickey Group at Colorado State University designed to provide students with a better understanding of the growth process of the nanoparticles and the relationship between solution color and nanoparticle size (*26*). We have also been working to further extend this experiment. Because the color changes observed from nanoparticle aggregation can be used for detection of oligonucleotide markers for disease (*27*), we are working to develop an activity that allows students to perform a forensic-type analysis for a specific strand of DNA based on the nonspecific binding of DNA to Au nanoparticles (*28, 29*), however, the delicate chemistry may preclude implementing this in large general chemistry laboratories.

During the 2007-2008 academic year, we intend to pilot a ferrofluid lab in several sections of the first semester general chemistry laboratory (*30*). The procedure will be rewritten to emphasize stoichiometric concepts including limiting reagent and yield. Eventually, we anticipate using this lab in all first semester sections and the modified version of the *Color My Nanoworld* (*17*) in all second semester sections.

Based on our experiences with incorporating nanoscience into the general chemistry lab, we have come to the conclusion that students will develop a more lasting and meaningful understanding of nanoscience when laboratory activities directly relate to concepts covered in class and connect to applications with which students have some familiarity. Because of the limited number of such existing activities, we are working to develop activities appropriate for the general chemistry laboratory. Our underlying goal for a successful laboratory is to use examples from nanoscience to illustrate fundamental general chemistry principles.

Organic Chemistry Lecture and Laboratory

Students majoring in biology or chemistry, or completing preparatory coursework for certain health professions, enroll in Organic Chemistry (CHEM 341/342). At JMU, organic is taught as a two semester lecture sequence; each lecture course is three credit hours. Chemistry majors enroll in a two semester integrated inorganic/organic laboratory sequence (CHEM 387L/388L) in which each course is worth two-credit hours. A two-credit laboratory, CHEM 346L, is taught concurrently with the second semester lecture (CHEM 342) for non-majors. Approximately 300 students per year begin the organic sequence, which is currently taught in two lecture sections of approximately 150 students, meeting three days per week. Laboratory sections (capped at 24 students) meet for one four-hour period each week. The enrollments in both general and organic chemistry courses have been rising in recent years and these trends are expected to continue as the university grows.

Our students are exposed to several nanoscience concepts in the organic lecture sequence. During the chapter on aromaticity, students study the structural and physical attributes of fullerenes and carbon nanotubes. Later, when learning about the properties and reactions of carboxylic acids and related derivatives, they are introduced to the concept of the micelle (*31*) – a self-assembled supramolecular aggregate formed when the salt of a carboxylic acid bearing a long chain (an amphiphile) is added to water. Both of these concepts are actually included in each of the corresponding chapters in the students' textbook, Carey's *Organic Chemistry* (*32*). In fact, Carey has consistently displayed an image of a buckyball or a related material on the cover of each edition since the first edition in 1987 (two years after the discovery of fullerenes). The most recent (7th) edition features an intriguing image of a nanotube covalently modified with ferrocene moieties. A glance at the cover art is enough to spark some students' interest in nanoscience and its connections to, in this case, organic chemistry.

Several of the organic instructors often demonstrate a property of surfactants (a contraction of "surface active agents" – a term given to amphiphiles that tend to aggregate at the interface of two dissimilar phases). A petri dish containing a small amount of water (~ 5 mm deep) is placed on an overhead projector, onto which a small amount of ground black pepper is poured. The instructor then dips his/her finger into a dilute soap (surfactant) solution, and then gently touches the center of the petri dish. The surfactant quickly spreads across the surface of the air-water interface, as clearly visualized by the pepper rushing to the edges of the dish. This simple in-class demonstration typically leads into a discussion of surfactant behavior, including the formation of monolayers, bilayers, micelles, and the relationship between structure and aggregate formation.

The concept of surfactant self-assembly is revisited in the organic laboratory. A manuscript detailing this experiment has been submitted elsewhere (*33*), and is briefly summarized here. In a single laboratory session, students synthesize and investigate the colloidal properties of a gemini surfactant – an amphiphilic molecule with two non-polar tails and two polar heads connected by a spacer (see Figure 2a) (*34, 35*). The synthesis (a double Menshutkin reaction) and purification are straightforward and reproducible by organic lab students.

We have optimized three different sets of reaction conditions to make the synthesis amenable to a number of laboratory settings. Reasonable yields can routinely be obtained in 30 minutes (using a microwave reactor), one hour (under reflux), or one week (at room temperature in a sealed flask). The product is precipitated from the crude reaction mixture by addition of acetone and then separation by vacuum filtration.

Students then prepare three aqueous samples – one below the critical micelle concentration (or *cmc*, the concentration required for micelles to form), one above cmc, and one significantly above cmc. The addition of a fluorescent probe (with a visible emission that is sensitive to the polarity of its environment) presents students with a visible cue of the formation of micelles using a hand-held ultraviolet lamp. At low surfactant concentration, the luminescent probe is in a polar environment (water), while above cmc it becomes trapped within the non-polar interior of the micelle. Comparison of the color and intensity of samples below and above cmc provides evidence for the notable difference of environment in the immediate vicinity of the probe (see Figure 2 b-c). The third sample becomes quite viscous, which provides evidence for the presence of elongated worm-like micelles that become entangled at high concentration (*33*).

The use of this lab has met several objectives. First, it deliberately introduces nanoscience into the organic laboratory in that students prepare and learn about the self-assembly of micelles – aggregates that are approximately 2 nm wide. Second, it provides an opportunity for students to not only perform a synthesis, but to also study the properties of the new molecule that they have prepared (*36*). Third, and perhaps most importantly, it provides molecules to make a connection between this lab and other upper level labs. The material synthesized in this lab is used again in both the physical chemistry and *Science*

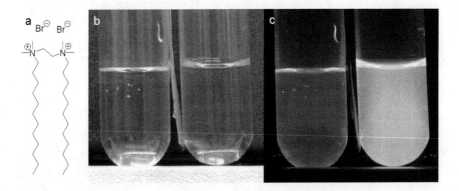

Figure 2. (a) Structure of one of the gemini amphiphile used in the laboratories. (b-c) Aqueous samples of a gemini amphiphile containing a fluorescent probe below (left tube) and above (right tube) critical micelle concentration, with the room lights turned on (b) and under UV-irradiation (c).

of the Small labs, where students complete a series of cmc studies using fluorescence spectroscopy, conductivity, surface tensiometry and diffusion-ordered nuclear magnetic resonance spectroscopy (DOSY NMR), *vide infra*.

We are also working on further developing this laboratory for the *Science of the Small* course by exchanging the two bromide counterions for a bisanionic chiral tartrate (D or L) (*37-39*). This ion exchange yields a product capable of forming elongated helical supramolecular fibers thus forming a thermoreversible gel in both aqueous and organic environments. The degree of twist of the fibers is controlled by the enantiomeric excess of the chiral counterion, thus providing external control of the supramolecular aggregation on the nanoscale.

Introduction to Materials Science

Chemistry and physics students may choose to take an elective lecture course that was developed jointly between the chemistry and physics departments at JMU, entitled "*An Introduction to Materials Science.*" The course is cross-listed in both departments (CHEM 275, PHYS 275) and is a required course in the materials science minor (listed as MATS 275). The course fulfills elective requirements in both majors. It is a three credit lecture course that has been team taught each year by a faculty member in both chemistry and physics. There are typically 8 – 20 students enrolled in this course.

MATS 275 emphasizes an atomistic understanding of the structure and properties of materials since it is targeted primarily at chemistry and physics majors. Topics covered in this class are standard for introductory materials science for engineers: crystal structure, X-ray diffraction, defects in solids, diffusion, phase diagrams, energy bands, mechanical, electrical and optical properties of solids. Over the last three years, the focus of this class has shifted to emphasize nanoscale phenomena. Structure-property relationships of materials such as carbon-nanotubes have been addressed in order to understand the mechanical, electrical, and optical properties of materials at the nanoscale. Also, nanoscale fabrication using self-assembly and templating as well as top-down engineering issues have been addressed in lectures on micro/nanofabrication. Each time that the course has been taught, examples from the current literature have been used to illustrate the "standard" topics in an introductory materials science course.

Physical Chemistry Laboratory

Nanoscience topics have been deliberately introduced into the the physical chemistry laboratory for ACS-track students (CHEM 438L) which covers quantum mechanics and spectroscopy. This course is required for all ACS-Certified degree students. These topics have been developed by replacing or enhancing existing labs with ones that introduce the same concepts using nanoscale examples. This is particularly easy to do because the quantum size effects observed in nanoscale materials are most easily addressed in physical chemistry. An experiment demonstrating quantum size effects on semiconductor luminescence was introduced into the CHEM 438L course nearly ten years ago when we developed a laboratory exploring the luminescence properties of porous silicon (p-Si) (*40*). This experiment was placed immediately following the traditional absorption spectroscopy experiment with conjugated cyanine dye molecules demonstrating the "particle-in-a-box" concept (*41-43*), and serves to directly reinforce the concept of the size-dependent luminescent properties of conjugated molecules with a solid state analog. The experiment was well-received by students as they had never had the opportunity to use silicon wafers and had generally performed few solid state chemistry experiments. In an additional module, students were introduced to AFM to observe the surface changes in the p-Si after electrochemical etching.

While effective in demonstrating semiconductor luminescence, surface science topics, and reinforcing the concept of quantum confinement in semiconductor materials, we decided to develop an alternative strategy to study semiconductor luminescence because this experiment required hazardous electrochemical etching solutions (HF:EtOH). In the literature there are many reports of the synthesis of direct band-gap II-VI semiconductor quantum dots (QD) with exquisite QD size control that produce a rainbow of luminescent colors (*44-46*). This chemistry has been reported in the chemistry education literature as well (*47, 48*). Because the synthesis requires dimethyl cadmium to be refluxed at a high temperature, there are both health and disposal concerns associated with developing this as a laboratory experiment. We have developed a room temperature synthesis that uses a MeOH:H_2O solvent system and a branched polymer as the passivating agent (*49*). While the quantum yield for QDs prepared by this method is smaller than that for QD prepared from $Cd(CH_3)_2$, the reagents are much less toxic. Recently, others have also reported synthesis procedures that are more amenable to use in an undergraduate teaching laboratory (*50-52*).

Two additional experiments have been recently added; both demonstrate nanoscience topics and reinforce concepts of quantum chemistry and spectroscopy. The first is an extension of the gemini amphiphile experiment that we use in organic chemistry (*33*). Dissolved fluorescent dyes are sensitive to the polarity of the solution, and thus both the intensity and wavelength of the fluorescence are affected by their environment. Students used material previously prepared in the organic teaching laboratories, and studied the fluorescence properties of the molecule pyrene over a range of amphiphilic concentrations. Careful examination of fluorescence spectra reveals subtle but reproducible differences in the relative intensities of the first and third emission peaks (I_1 and I_3) above and below the cmc, enabling students to indirectly measure the concentration at which micelles start to form (*33*). Figure 3 shows sample data from this experiment.

Finally physical chemistry laboratory students were introduced to scanning probe microscopy (SPM) and molecular manipulation on surfaces through the use of both AFM and scanning tunneling microscopy (STM) and a demonstration of a NanoManipulator haptic device (*53*). The development of the SPM in the late 1980s has helped to usher in many of the advances in nanoscale manipulation of materials on solid surfaces. AFM in particular has enabled atomic (or nearly atomic) resolution of many classes of materials ranging from semiconductors to biomaterials (*54*). The NanoManipulator instrument is a haptic device that interfaces to a SPM instrument and is available through 3rd Tech, Inc. It currently interfaces to three different commercial AFMs; the Veeco

Instruments Explorer, Nanotec Electronica, and Asylum Reseach MFP-3D. Students were asked to compare the surface roughness of three common papers; namely glossy photo paper, standard copy paper and an uncirculated dollar bill via AFM. They were also asked to measure the average width, length and distance between data encoded on a compact disc. Using the STM, students were able to obtain lattice images of highly ordered pyrolytic graphite (HOPG),

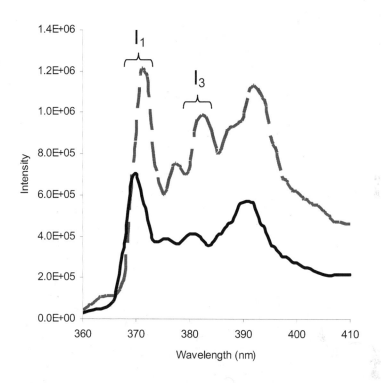

Figure 3. Fluorescence emission spectra of pyrene in aqueous preparations of a gemini amphiphile above (broken line) and below (solid line) the critical micelle concentration. Note that the I1/I3 ratio changes from 1.23 (above cmc) to 1.72 (below cmc).

measure the average bond lengths and distances, and compare it to a model. Sample student data of several of these substrates are shown in Figure 4.

The NanoManipulator tool enabled students to "feel" the surface of a carbon nanotube on a graphite surface and DNA from a ruptured virus on a mica surface (55). Both samples were pre-packaged with the NanoManipulator software. The NanoManipulator hardware and software were originally developed for laboratory-based microscopes and are thus limited to a location in a microscopy laboratory. Researchers at the University of North Carolina developed the capability to operate the NanoManipulation haptic device by connecting remotely to the AFM via the internet for educational demonstrations. They have used this capability to demonstrate "feeling" or "manipulating" objects at the nanometer-scale in real-time to college and high school level audiences (55). The primary disadvantage of this approach is that there needs to be a skilled

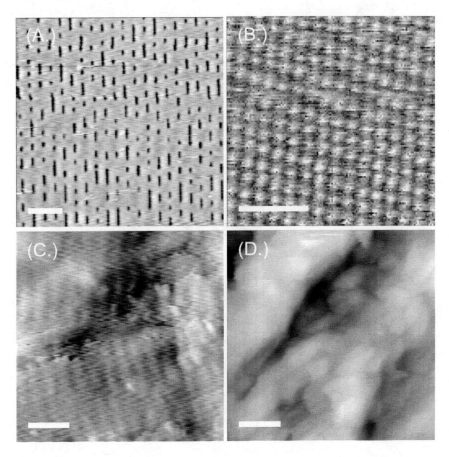

Figure 4. Sample scanning probe microscopy images obtained by students in Physical Chemistry Laboratory. (A.) 30 × 30 μm AFM image of the surface of a compact disc. Scale bar = 5 μm, (B.) 3 × 3 nm STM image of highly ordered pyrolytic graphite in constant current mode. Scale bar = 1 nm. (C. and D.) 5 × 5 μm images of standard copy paper and U. S. currency, respectively. Scale bar = 1 μm. Z-range is 170 nm and 887 nn for (C.) and (D.), respectively.

AFM operator performing the demonstration at the remote site and another skilled operator at the "home-base" of the instrument to ensure everything is working properly. We are currently developing an interface between the easyScan AFM and the NanoManipulator enabling a completely portable system. The computer, AFM controller, haptic interface, portable LCD projector and all necessary cables fit into a rugged instrument case slightly larger than an airplane

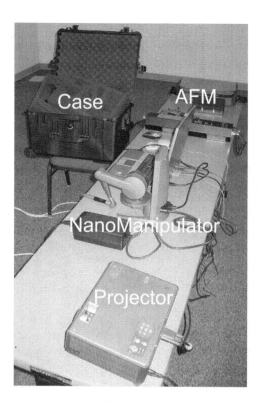

Figure 5. Portable AFM / NanoManipulator hardware consisting of Nanosurf easyScan 2 AFM, 3rd Tech NanoManipulator haptic control, portable LCD projector, associated power supplies/cables and carrying case.

carry-on bag as shown in Figure 5. We have successfully developed the software to interface the NanoManipulator and the easyScan AFM so that it can operate in a "real-time feel" mode, and are currently working with the software developers at Nanosurf to enable the "point control" or active mode.

Literature and Seminar I and II

All chemistry students are required to take a pair of one credit lecture courses to learn how to search, read, interpret, analyze, and present the scientific literature. These courses, Literature and Seminar I/II (CHEM 481/482) focus nearly exclusively on these skills. During the first semester, the science librarian introduces students to JMU's electronic research tools for the first half of the semester while a faculty member works with students on reading, interpretation, and analysis of current papers during the second half of the semester. During the second semester, students complete several projects including a unit on ethics, a debate on science in the news and a review paper and presentation on a "hot topic" in science.

The Literature and Seminar courses provide an ideal venue for the incorporation of nanoscience in the curriculum because a skill set, rather than chemistry content, is covered. A faculty member with some expertise or interest in nanoscience can select current papers that fall under the purview of nanoscience. A faculty member without this expertise can use this as an opportunity to learn more about this or another field. For example, in the Fall 2006 offering of the class, the faculty member teaching this first semester course selected only nanoscience papers to introduce students to the process of reading scientific papers. For the initial reading assignment, students discussed a *JACS* paper where modified magnetite particles were used to remove uranyl ions from blood (56). For their second assignment, students had to write a memo discussing the use of a newly synthesized molecular sieve for use as a catalyst (57). In the final literature-based assignment, students were introduced to the idea that sometimes conflicting models exist to explain a chemical phenomenon by reading papers that discussed using DNA as a molecular nanowire (58, 59). The semester concluded with a presentation of notable research papers from the past year to get students to think about topics for their review paper and presentation during the second semester. Frequently, this talk has provided the seeds for student review ideas. During the subsequent semester, the topics that were discussed were biased towards nanoscience: one-quarter of the class chose to complete their second semester review papers and talks on topics that dealt with aspects of nanoscience.

New Course Development

Two new classes have been developed to support nanoscience awareness at JMU. The first is a general education class entitled "*Physics, Chemistry and the Human Experience*" (GSCI 101). The second is an upper-division elective lecture-laboratory course primarily designed for science and technology majors entitled, "*Science of the Small: An Introduction to the Nanoworld*" (CHEM 480P). Each course will be discussed in further detail.

Physics, Chemistry, and the Human Experience

As part of the requirements for graduation, all JMU students must complete general education courses. There are five clusters of courses, spanning the academic disciplines, including one dedicated to math and science in the natural world. The description of this cluster from JMUs general education website is provided.

> *Scientific investigations into the natural world use analytical methods to evaluate evidence, build and test models based on that evidence, and develop theories. Mathematical studies of form and pattern can create a language that assists in these investigations. In this ten-to-eleven*

credit hour cluster, students are provided with the opportunity to develop problem-solving skills in science and mathematics at the college level. Students will be introduced to a substantial body of scientific facts, concepts, models, and theories and will also gain experience in using basic mathematics to obtain knowledge about the natural world. Each package is multidisciplinary and interdisciplinary, thereby demonstrating boundaries and connections among mathematics, the sciences, and other aspects of culture.

One of the core science courses in this cluster is GCSI 101 – *Physics, Chemistry and the Human Experience*. Professors have the freedom to design sections of GSCI 101 to spotlight any topic within the broad realm suggested by the course's title. In the past, the science of medical technology, environmental science and the history of materials (among others) have been used as themes to address the course goals.

In the spring of 2004, one of the authors developed a GSCI 101 section with a focus on nanotechnology. The students in the course learn basic chemistry and physics topics in the context of the fundamental principles, properties and technology in the nanoscale size regime. While the course concentrates on the science of nanotechnology, it also provides an ideal platform to discuss a broad range of topics including the interdisciplinary nature of science, science funding, ethics, politics and culture. Most importantly, it creates informed citizens by exposing non-science majors to the intricacies of nanotechnology in a way that they might not be able to glean from the popular press.

The required texts for the course are: *Designing the Molecular World. Chemistry at the Frontier (60)*, *Nanotechnology. A Gentle Introduction to the Next Big Idea (61)*, and *The Pleasure of Finding Things Out (62)*. Topics addressed include both chemistry and fundamentals in math and physical science including including unit conversions, relative size, atoms, molecules, bonding, VSPER, intermolecular forces, water, DNA, the electromagnetic spectrum, and nanoscience-oriented topics such as buckyball/nanotubes, molecular switches, host-guest chemistry, self-assembly, liquid crystals, smart materials, lithography, and scanning probe microscopies. The portable AFM was also demonstrated in this class, allowing students to operate this modern instrumentation.

At the end of the semester, students prepared and presented "Nanopresentations." Teams of two students each were responsible for researching and presenting information on a company that is involved in nanotechnology research and/or production. They considered both companies that are entirely focused on nanotech, as well as those that are more broadly focused, but use nanotechnology for one or more applications or products. Students took on the role of company representatives trying to convince the class to invest in their company. Students generally enjoyed learning about nanotechnology; a few even sent the instructor periodic correspondence for years after taking the course whenever they heard about nanotechnology on the news.

Notably, while this section of GSCI 101 was developed with a core nanotechnology focus, at least three other GSCI 101 sections included brief nanotechnology units (often including an AFM demonstration), thus significantly increasing the affected audience.

Science of the Small: An Introduction to the Nanoworld

The *Science of the Small* (SoS) course was designed as a four credit elective lecture-laboratory class targeted to upper division science and technology majors. Prerequisites for the course were one semester of general chemistry and two semesters of general physics. The initial offering of this course was taught in Spring 2007 and six students were enrolled. In the initial class, five students were chemistry majors and one was an integrated science and technology and physics double major. The class was team taught by the three authors and the lecture topics were broadly divided into three major themes. In order of presentation these were: top-down engineering, bottom-up engineering and self-assembly, and low dimensional systems. The lecture class met twice a week for 75 minutes and each major theme was allotted nine to ten lectures including an exam for each major topic. Table III lists the broad lecture topics for the course in sequential order. Three guest lectures were also presented: (1) Dr. Scott Paulson from the JMU Department of Physics and Astronomy discussed carbon-based nanomaterials during the top-down engineering section, (2) Dr. Michal Sabat from the University of Virginia Department of Chemistry presented a lecture on nanomachines during the self-assembly section, and (3) Dr. Lisa Fridersdorf, representing Virginia's legislative efforts from the Joint Commission on Technology and Science Advisory Committee on Nanotechnology (*63*), spoke about policy issues and corporate investment in nanotechnology primarily within the state of Virginia.

The course also included a 3.25-hour laboratory each week. The laboratory portion of the class featured the same broad highlighted themes—top-down, bottom-up, and low dimensional systems—in four "modules." Each module was a multi-week experiment in which students worked as research teams of three students to conduct the experiments and produce a formal laboratory report. The four modules are also included in Table III.

When designing the *Science of the Small* course, we aimed to equip our students with a firm understanding of the core theoretical concepts of nanoscience such as quantum effects, thermodynamics of self-assembly, band vs. molecular picture of energy levels, while giving them a broad view of the research and technology included in this burgeoning field. Thus, after providing appropriate background information, we focused on case studies of recently published material from diverse perspectives, demonstrating different approaches to solving modern scientific problems.

One of the challenges facing highly interdisciplinary elective courses with relatively few prerequisites is that there is often a broad distribution of backgrounds of the participating students. In this particular course, three of the students were juniors and three were seniors. One had already taken the introductory materials science course (MATS 275) as well as several engineering-type courses, but was concurrently enrolled in General Chemistry II. The chemistry majors had no exposure to materials science yet had a strong background in chemistry. In addition, several of the students had already had Physical Chemistry II (quantum mechanics), while others had no exposure to quantum concepts apart from the most basic treatment provided in general and inorganic chemistry. These issues will only be exacerbated if future iterations of the class include biology majors who generally have no exposure to materials science and engineering ideas, minimal exposure to physics, and no meaningful experience with quantum mechanics.

When we prepared the course outline, we specifically designed the lecture topics to correspond to the laboratory modules, and when we divided up the responsibilities for the course, each faculty member was assigned according to his or her own background and field of expertise or interests. In the laboratory section of the course, Augustine developed and taught micro/nanofabrication and characterization, Caran led the organic synthesis and molecular self-assembly section and Reisner headed up the low dimensional systems and DNA passivated Au nanoparticle synthesis and characaterization. Pedagogically it was convenient to have the topical content of the lecture and laboratory correspond. Additionally, these topics naturally progress from the micro to the nano scale and literally illustrate the top-down vs. bottom-up concept. This is aesthetically appealing. It however, was challenging for each of the faculty to teach the lecture and laboratory simultaneously. This was particularly true with the laboratories since none of the experiments were "off the shelf." Many required delicate and often finicky equipment, sensitive reagents, and relatively sophisticated characterization and spectroscopic techniques. This issue of scheduling both the lecture and the laboratory simultaneously will become less challenging in subsequent offerings of the class as the lectures will be more fully developed. However, this should be taken into consideration if one is interested in developing a similar course.

One of the students enrolled in the class also served as a teaching assistant for the laboratory. She prepared necessary reagents, located and made available the appropriate supplies, and provided support before, during and after the experiments. Importantly, she also worked with the faculty to develop and optimize the organic synthesis and self-assembly modules. Having a student TA familiar with one of the modules contributed to the success of the laboratory portion of the SoS class, and is recommended for the first iteration of a similar course.

Table III. Topic Schedule of Science of the Small Course: Spring 2007

Lecture Topics	Laboratory Experiments
Layers on Surfaces: Top-Down Engineering	
Introduction to Nanotechnology	Intro to facilities / Laboratory protocol / safety / lab tour
The Quantum World	(Module #1): Deposition and characterization of thin films (metals and polymers including ellipsometry, XRD, AFM, SEM)
Properties of Materials	(Module #2): Microfabrication of Si master / PDMS stamp (*64*), and μ-CP and edge spreading lithography (*65*) and characterization.
Case Study: Carbon-Based Nanomaterials (Guest lecture)	
Tools to Visualize/Manipulate Matter	
Soft Materials: Bottom-Up Engineering / Self Assembly	
Supramolecular Chemistry Introduction	(Module #3): Synthesis and purification of biscationic gemini amphiphiles / ^{1}H and ^{13}C NMR characterization. Characterization of cmc using surface tensiometry, conductivity, fluorimetry, and diffusion rate (DOSY NMR) (*33*)
Supramolecular Chemistry Literature Review	
Introduction to Colloid Chemistry	
Molecular Machines (Guest Lecture)	
Literature Review Case Studies: Molecular Boromean Rings, Nanobiolithography and Nanostructured biomaterials	
Low Dimensional Systems: Nanoparticles and Nanostructures	
Literature Review Case Studies: Oligonucleotide Recognition Noble Metal Nanostructures Semiconducting Nanostructures Nanoporous Materials	(Module #4): Research project on oligonucleotide recognition using Au nanoparticles (*28, 29*).
Nanoscience and Environmental Impacts	
Commercial Opportunities for Nanoscience in Virginia and Beyond (Guest Lecture)	
Student Presentations: Proposal for a Nanobusiness	

Lessons Learned from the Systematic Incorporation of Nanoscience into the Undergraduate Curriculum

"Top-Down" Nanoscience Curriculum Develoment

While the original goal of this project was to seed nanoscience throughout the curriculum, an unexpected result of the *Science of the Small* course development has been a catalytic seeding process of the entire curriculum from the "top-down." The development of laboratory activities and lecture materials for this capstone course has helped to focus and direct our efforts to initiate changes at all levels of the chemistry curriculum. In the process of developing new laboratory experiments, we were able to identify other experiments that had the potential to be used in the required introductory laboratories in our curriculum. This is a work that is ongoing, but the task of developing an entire suite of experiments for a highly interdisciplinary course has led to ideas for many "spin-off" laboratory experiences in lower-division courses.

A critical enabler of this subtle shift has been having several faculty members each with different specialties developing the course and closely working together. Since the authors' teaching and research areas span a variety of chemistry disciplines – specifically, physical / materials, organic / supramolecular, and inorganic / materials – our teaching responsibilities span freshman through senior offerings in general, organic, inorganic, physical, and materials chemistry. The SoS class has caused us to look critically at how and where nanoscience topics can be introduced throughout the curriculum. There are subtle and ongoing changes being implemented at all levels of the chemistry major. Further, since the course was offered in Spring 2007, the catalytic effect is just now being felt. All three of the authors have begun to identify even more areas of intersection with other traditional chemistry courses and are more comfortable making connections between core areas by using nanoscience examples. In addition, we are now able to provide other faculty members in the Departments of Chemistry and Biochemistry and Physics and Astronomy with materials that they can use to implement more nanoscience actvities into their courses. For example, in teaching this course, we have developed lecture materials that can be used by biochemistry faculty when discussing imaging techniques.

Perhaps the most far-reaching and significant potential impact of this project is that it has caused faculty from other interdisciplinary fields represented in the department (biochemistry and biotechnology, environmental science, etc.) to think about their fields in a more holistic manner and to seed their interdisciplinary interests across the curriculum. It has even helped to catalyze and crystallize some of the discussions about future faculty hires and the expertise that we should be seeking. The philosophy of implementing evolutionary changes to the chemistry curriculum that has been used in developing this project has become a model to be emulated across the department.

For example, the subtle catalytic shift prompted by seeding interdisciplinary topics throughout the core of the chemistry degree has led to discussions about thoroughly reworking our general chemistry laboratory program so that a majority of the second semester laboratory experiments center around interdisciplinary themes. The concepts traditionally taught in the course will remain the same, but the actual experiments will be modified or replaced. An example of this is in a quantitative absorption spectroscopy experiment, students would still be required to carefully produce a calibration curve, understand Beer's Law, and measure the unknown concentration of an analyte. However, rather than investigate a generic inorganic compound in water, the analyte might be a protein in a buffer solution. This system then naturally leads to a discussion about protein folding and denaturing, high molecular weight systems, and more sophisticated analysis methods for proteins. Our new labs will highlight the connections and applications of chemistry while continuing to cover the basic concepts needed in general chemistry. This example of an evolutionary change multiplied over dozens of experiments with themes in materials/nanoscience, enviromental/green chemistry, and biochemistry throughout the core curriculum begins to better prepare students for real-world research challenges and the interdisciplinary problems at the heart of modern chemical research.

By better understanding how nanoscience can be used as examples to support core disciplinary topics, the crux of the evolutionary approach, we now think differently about how interdisciplinary topics can be used in all courses. Rather than teaching a chemistry core with minimal interconnectedness and then offering specialized upper-division interdisciplinary elective courses, we seek to seed interdisciplinary topics throughout the core *and then* offer elective courses.

This strategy accomplishes at least four separate goals. Students may be more inclined to take an elective if they have been exposed to the material and find it interesting. Students taking electives are better prepared for the material. Even without taking elective courses, students are able to better appreciate how the core courses relate to one another and to disciplines outside of the major such as physics and biology. Most importantly, students graduating with an undergraduate degree in chemistry have a much better overall understanding of the interdisciplinary nature of chemistry as it is practiced by professional chemists in both industry and academia.

Laboratory Development: An Opportunity for Undergradute Research and Pre-Service Teacher Training

Identifying relevant and appropriate experiments for laboratory experiments in diverse courses is the primary challenge for this approach. An overarching theme in the experiments that we have identified as potentially useful in teaching laboratories is a desire to adapt experiments from the recent nanoscience literature. We have found that students respond favorably to experiments that have recently been published in the scientific literature, but as we have discussed previously, there are practical requirements that must be met for laboratory experiences at each level of the curriculum which often make the adaptation of recent literature difficult.

It takes skill and practice to adapt recent papers into meaningful classroom activities, and as such, provides an outstanding research opportunity. Over the past two years, we have worked with several undergraduate students to develop laboratory activities. We have found that these experiences are particularly good for traning pre-service teachers.

JMU was originally founded as a teachers college, and while the university has expanded in size and mission greatly since 1908, teacher training still remains a vital part of the mission of JMU and many aspiring teachers still consider JMU one of the top schools in Virginia for learning the craft of teaching. It is from this pool that we have sought to identify and hire summer research students to help develop appropriate laboratory experiments. Developing, modifying and adapting laboratory experiences that are appropriate for a high school or a middle school audience with a limited set of equipment and budget is an invaluable skill for an aspiring secondary school teacher. Unfortunately, these skills are rarely taught in any formal or informal setting in the undergraduate science or education curricula. This skill set is best realized in the research laboratory where one is constantly required to innovate, modify and adapt ideas to test a hypothesis. By hiring pre-service teachers to work on adapting cutting-edge science to a laboratory setting, students are learning the craft of research because, in general, recently published experiments require patience and skill to reproduce. They are also learning the craft of teaching experimental science because they are forced to identify which parts of the published work are amenable to being adapted to the undergraduate (or other) teaching laboratory. We have found that the development of laboratory activities provides an outstanding research experience for future secondary education teachers as they have an opportunity not only to experience the benefits of undergraduate research, but they also learn some of the pedagogy related to laboratory development.

Assessment of Nanoscience Experiences in the Undergraduate Curriculum

Over the course of this project, our assessment goals have evolved. Initially, we sought only to measure students' knowledge of nanoscience concepts. When we began this project, no validated instruments were available to assess nanoscience knowledge. Since this time, the ACS Exams Institute has developed a series of nanoscience questions that may be used with several exams (66). To this end, we developed an exam to test student knowledge, the Nanoscience Assessment Instrument (NAI). This exam is currently being administered to all graduating chemistry majors during February of their last year. It is also being administered as the final exam for the *Science of the Small* course.

The original version of this exam tested primarily content knowledge., i.e. does a student know a particular fact set. However, a cursory analysis of student performance on both this exam and the ACS Exams Institute's Full-Year General Chemistry Exam made us realize that we could not determine whether students' poor performance on the instrument was because of a lack of knowledge about nanoscience or a lack of general content knowledge. For this reason, we began the development of a paired exam, with a philosophy similar to that employed in the ACS Exams Institute's Paired General Chemistry

Exams. In our current version of the NAI, questions are divided into several categories: the atomic world; size and scaling; synthesis, self-assembly, and reactivity; structure; properties of materials; science, technology, and society; and characterization. Items on the instrument address not only content knowledge, comprehension, and application, but also analysis and synthesis as classified by Bloom's taxonomy (*67*). By looking at the paired questions on this exam, we are refining the NAI to keep items that provide good discrimination while removing items that provide no information. Over the course of this project, our ideas have evolved in a way that we now believe that providing more experiences in nanoscience will help our students develop a better molecular level picture which is essential for understanding chemistry. We have begun the process of data collection to determine the validity of this hypothesis.

One area which we have not assessed is student attitudes towards nanoscience. As educators, we are not only responsible for helping students develop their content knowledge and critical thinking skills, but also their understanding of how science impacts society. By incorporating nanoscience into our curriculum, we have the potential to affect student attitudes towards nanoscience. To understand the role that our approach has, we will be assessing attitudes for a number of populations including students with no nanoscience experience, general science students, general chemistry students, students taking upper-division service courses in chemistry, majors, and students taking a course in nanoscience. Work is underway on the development of an instrument to assess student attitudes towards nanotechnology as they progress through the curriculum. Our instrument will be based upon survey questions already tested in other studies (*68-70*).

Conclusions

Nanoscience has become a catalyst for change within the Department of Chemistry and Biochemistry at James Madison University over the past four years. What began simply as an idea for a proposal to introduce an emerging subdiscipline of chemistry into the current curriculum has developed into a new model of thinking about how to integrate interdisciplinary themes throughout the undergraduate chemistry curriculum. By better understanding and addressing the hurdles facing these changes; expertise, class size, varying pedagogies—we have been able to make changes supported by the faculty. This program has helped us to develop a model of seeding interdisciplinary topics liberally throughout the core and has helped the faculty to consider how these core courses relate to one another. An unanticipated result of the *Science of the Small* course has been for faculty specializing in other interdisciplinary areas of modern chemistry to consider how their fields might be seeded throughout the curriculum in laboratory and lecture examples in a similar fashion. We are now seriously considering how many of the time-worn experiments in general chemistry laboratory can be replaced with more modern, interdisciplinary experiments that will correlate to laboratory and/or lecture topics. This process has forced us to evaluate our curriculum from the freshman to senior level to provide an academically rigorous, robust and relevant curriculum for our chemistry majors as they enter the workforce or graduate programs in the highly interdisciplinary world of science and technology. Perhaps most importantly, we have placed nanoscience into the volcabulary of a wide range of students including both science and non-science majors.

Acknowledgements

The authors would like to thank the JMU undergraduate student collaborators who have aided in the development of this project: Stephanie Torcivia, Patrick Turner, Jon Wyrick, Stephen Winward, John DeJarnette, and John Magnotti. We would also like to acknowledge faculty efforts in developing laboratory and lecture topics including Dr. Debra Mohler (JMU Chemistry and Biochemistry), Drs. Scott Paulson and Chris Hughes (JMU Physics and Astronomy), Dr. Michal Sabat (UVA Chemistry) Dr. Lisa Friedersdorf (UVA Materials Science and Engineering), and Mrs. Kathryn Harrell and Ms. Jennifer Cunningham (Charlottesville High School, VA). We also thank the good patience of the first *Science of the Small* class who were unapologetically called upon to be guinea pigs for many of the lecture and laboratory experiments. Their feedback and honesty has already given ideas to improve the next iteration of the class. This work was primarily funded through the National Science Foundation Nanoscience Undergraduate Education program (NSF-NUE CHE-0532451). Additional funding was provided through the NSF Research Experiences for Undergraduates programs in Chemistry (NSF-REU CHE-035807), the Department of Defense ASSURE program (NSF DMR-0353773), the NSF Research at Undergraduate Institions program (NSF-RUI DMR-0405345), and the NSF Course Curriculum and Laobroatory Instruction program (NSF-CCLI DUE-0618829).

References

1. Rosei, F. *J. Phys. Cond. Matt.*, **2004**, *16*, 1373-1436.
2. Braun, E.; Keren, K. *Adv. Phys.*, **2004**, *53*, 441-496.
3. Roduner, E. *Chem. Soc. Rev.*, **2006**, *35*, 583-592.
4. Weiss, R., *Washington Post*; Washington, D.C., Feb. 1, 2004, A.01.
5. Weiss, R., *Washington Post*; Washington, D.C., Mar. 28, 2005, A.06.
6. Kahn, J. *National Geographic* **2006**; *209*, 98-119.
7. *Consumer Reports* **July 2007**.
8. Undergraduate Origins of Recent (1991-95) Science and Engineering Doctorate Recipients, **2006**, http://www.nsf.gov/statistics/nsf96334/nsf96334.pdf, *Last accessed August 2007*
9. Kittredge, K. W.; Russell, L. E.; Leopold, M. C. *Chem. Educ.*, **2007**, *12*, 155-158.
10. Amenta, D. S.; Mosbo, J. A. *J. Chem. Educ.*, **1994**, *71*, 661-664.

11. Light, G.; Swarat, S.; Park, E. J.; Drane, D.; Tevaarwerk, E.; Mason, T. *1st Int. Conf. Res. Eng. Educ.*; Froyd, J., Ed.; Amer. Soc. Eng. Educ.: Honolulu, HI, **2007**
12. Nanopedia, **2007**, http://nanopedia.case.edu/NWPage.php?page=nanoscience, *Last accessed August 2007*
13. Furton, K. G.; Norelus, A. *J. Chem. Educ.*, **1993**, *70*, 254-257.
14. Berger, P.; Adelman, N. B.; Beckman, K. J.; Campbell, D. J.; Ellis, A. B.; Lisensky, G. C. *J. Chem. Educ.*, **1999**, *76*, 943-948.
15. Campbell, D. J.; Olson, J. A.; Calderon, C. E.; Doolan, P. W.; Mengelt, E. A.; Ellis, A. B.; Lisensky, G. C. *J. Chem. Educ.*, **1999**, *76*, 1205-1211.
16. Campbell, D. J.; Beckman, K. J.; Calderon, C. E.; Doolan, P. W.; Ottosen, R. M.; Ellis, A. B.; Lisensky, G. C. *J. Chem. Educ.*, **1999**, *76*, 537-541.
17. McFarland, A. D.; Haynes, C. L.; Mirkin, C. A.; Van Duyne, R. P.; Godwin, H. A. *J. Chem. Educ.*, **2004**, *81*, 544A-544B.
18. Haynes, C. L.; McFarland, A. D.; Van Duyne, R. P.; Godwin, H. A. *J. Chem. Educ.*, **2005**, *82*, 768A-768B.
19. Olsen, K.; Hardin, L.; McGovern, J.; Manning, T. J.; Phillips, D.; Ayers, T.; Duncan, M. A. *Chem. Educ.*, **2005**, *10*, 260-264.
20. Manning, T.; Bennett, J.; Hardin, L.; Harris, K. *Chem. Educ.*, **2005**, *10*, 371-377.
21. Stuber, L. M.; Rachford, E. M.; Jordan, C. S.; Mitchell, S. J.; Tabron, C.; Manning, T. J. *Chem. Educ.*, **2005**, *10*, 204-207.
22. Richardson, A.; Janiec, A.; Chan, B. C.; Crouch, R. D. *Chem. Educ.*, **2006**, *11*, 331-333.
23. Solomon, S. D.; Bahadory, M.; Jeyarajasingam, A. V.; Rutkowsky, S. A.; Boritz, C.; Mulfinger, L. *J. Chem. Educ.*, **2007**, *84*, 322-325.
24. Winkelmann, K.; Noviello, T.; Brooks, S. *J. Chem. Educ.*, **2007**, *84*, 709-710.
25. MRSEC Interdisciplinary Education Group Video Lab Manual, **2007**, http://mrsec.wisc.edu/Edetc/nanolab/index.htm, *Last accessed August 2007*
26. Rickey, D., Personal communication.**2007**
27. Thaxton, C. S.; Rosi, N. L.; Mirkin, C. A. *MRS Bull.*, **2005**, *30*, 376-380.
28. Li, H.; Rothberg, L. J. *J. Am. Chem. Soc.*, **2004**, *126*, 10958-10961.
29. Li, H.; Rothberg, L. J. *Proc. Nat. Acad. Sci.*, **2004**, *101*, 14036-14039.
30. Bentley, A. K.; Farhoud, M.; Ellis, A. B.; Lisensky, G. C.; Nickel, A. N.; Crone, W. C. *J. Chem. Educ.*, **2005**, *82*, 765-768.
31. Menger, F. M.; Zana, R.; Lindeman, B. *J. Chem. Educ.*, **1998**, *75*, 115.
32. *Organic Chemistry, 7th Ed.;* Carey, F. A.; McGraw-Hill Science/Engineering/Math: New York, 2008.
33. Torcivia, S.; Caran, K. L. *J. Chem. Educ.*, **submitted 2007**.
34. Menger, F. M.; Keiper, J. S.; Mbadugha, B. N. A.; Caran, K. L.; Romsted, L. S. *Langmuir*, **2000**, *16*, 9095-9098.
35. Menger, F. M.; Keiper, J. S. *Angew. Chem. Int. Ed.*, **2000**, *39*, 1906-1920.
36. Horowitz, G. *J. Chem. Educ.*, **2007**, *84*, 346-353.
37. Oda, R.; Huc, I.; Candau, S. J. *Angew. Chem. Int. Ed.*, **1998**, *37*, 2689-2691.
38. Oda, R.; Huc, I.; Schmutz, M.; Candau, S. J.; MacKintosh, F. C. *Nature (London)*, **1999**, *399*, 566-569.

39. Berthier, D.; Buffeteau, T.; Leger, J.-M.; Oda, R.; Huc, I. *J. Am. Chem. Soc.*, **2002**, *124*, 13486-13494.
40. Lasher, D. P.; DeGraff, B. A.; Augustine, B. H. *J. Chem. Educ.*, **2000**, *77*, 1201-1203.
41. Herzfeld, K. F.; Sklar, A. L. *Rev. Mod. Phys.*, **1943**, *14*, 294.
42. Sheppard, S. E.; Geddes, A. L. *J. Am. Chem. Soc.*, **1944**, *66*, 2003.
43. Kuhn, H. *J. Chem. Phys.*, **1949**, *17*, 1198.
44. Wang, Y.; Herron, N. *J. Phys. Chem.*, **1991**, *95*, 525-532.
45. Alivisatos, A. P. *Science (Washington, D.C.)*, **1996**, *271*, 933-937.
46. Murphy, C. J.; Coffer, J. L. *Appl. Spectr.*, **2002**, *56*, 16A-27A.
47. Nedelijkovic, J. M.; Patel, R. C.; Kaufman, P.; Joyce-Pruden, C.; O'Leary, N. *J. Chem. Educ.*, **1993**, *70*, 342-344.
48. Kippeny, T.; Swafford, L. A.; Rosenthal, S. J. *J. Chem. Educ.*, **2002**, *79*, 1094-1100.
49. Winkler, L. D.; Arceo, J. F.; Hughes, W. C.; DeGraff, B. A.; Augustine, B. H. *J. Chem. Educ.*, **2005**, *82*, 1700-1702.
50. Lynch, W. E.; Nivens, D. A.; Helmly, B. C.; Richardson, M.; Williams, R. R. *Chem. Educ.*, **2004**, *9*, 159-162.
51. Boatman, E. M.; Lisensky, G. C.; Nordell, K. J. *J. Chem. Educ.*, **2005**, *82*, 1697-1699.
52. Hutchins, B. M.; Morgan, T. T.; Ucak-Astarlioslu, M. G.; Williams, M. E. *J. Chem. Educ.*, **2007**, *84*, 1301-1303.
53. Falvo, M. R.; Clary, G.; Helser, A.; Paulson, S.; Taylor Ii, R. M.; Chi, V.; Brooks Jr, F. P.; Washburn, S.; Superfine, R. *Microsc. Microanal.*, **1998**, *4*, 504-512.
54. Giessibl, F. J.; Quate, C. F. *Phys. Today*, **2006**, *59*, 44-50.
55. Jones, M. G.; Andre, T.; Superfine, R.; Taylor, R. *J. Res. Sci. Teach.*, **2003**, *40*, 303-322.
56. Wang, L.; Yang, Z.; Gao, J.; Xu, H.; Zhang, B.; Zhang, X.; Bing Xu, B. *J. Am. Chem. Soc.*, **2006**, *128*, 13358-13359.
57. Corma, A.; Díaz-Cabañas, M.; Jordá, J. L.; Martínez, C.; Moliner, M. *Nature (London)*, **2006**, *443*, 842-845.
58. Fink, H.-W.; Schönenberger, C. *Nature (London)*, **1999**, *398*, 407-410.
59. Porath, D.; Bezryadin, A.; de Vries, S.; Deeker, C. *Nature (London)*, **2000**, *403*, 635-638.
60. Ball, P.; *Designing the Molecular World;* Princeton University Press: Princeton, NJ, 1996.
61. Ratner, M. A.; Ratner, D.; *Nanotechnology: A Gentle Introduction to the Next Big Idea;* Prentice Hall PTR: Upper Saddle River, NJ, 2002.
62. Feynman, R. P.; *The Pleasure of Finding Things Out: The Best Short Works of Richard P. Feynman;* Basic Books: New York, 2005.
63. Joint Committee on Technology and Science, **2007**, http://jcots.state.va.us/2007%20Content/advcom07.htm, *Last accessed August 2007*
64. Xia, Y. N.; Whitesides, G. M. *Ann. Rev. Mater. Sci.*, **1998**, *28*, 153-184.
65. Geissler, M.; McLellan, J. M.; Xia, Y. *Nano Lett.*, **2005**, *5*, 31-36.
66. Holme, T.; Nanoscience Items for Standardized Examinations in the Undergraduate Chemistry Curriculum. In *Nanoscale Science and*

Engineering Education; Sweeney, A. E; Seal, S., Eds.; American Scientific Publishers: Valencia, CA, **2008**, 241.
67. Bloom, B. S.; *Taxonomy of Educational Objectives;* Green: New York, 1956.
68. Bainbridge, W. S. *J. Nanopart. Res.*, **2002**, *4*, 561-570.
69. Currall, S. C.; King, E. B.; Lane, N.; Madera, J.; Turner, S. *Nature Nanotech.*, **2006**, *1*, 153-155.
70. Waldron, A. M.; Spencer, D.; Batt, C. A. *J. Nanopart. Res.*, **2006**, 569-575.

Chapter 4

Vertical Integration of Nanotechnology Education

Vesselin N. Shanov,* Yun Yeo-Heung, Leigh Smith, Shalyajit Jadhav, Thang B. Hoang, Andrew Gorton, Thomas Mantei, John Bickle, Ian Paputsky, Frank Gerner, Julie L. Burdick, Alexzandra Spatholt, Mitul Dadhania, Gautam Seth, Mark J. Schulz * and Suri S. Iyer,*

University of Cincinnati, Cincinnati, OH 45221, USA

We describe our experiences teaching nanoscale science and technology to students at various levels in the Cincinnati, Ohio area. Faculty members from different departments were involved in this project. Teaching was done in a top down manner where faculty and students taught at different levels based on their training. We observed our approach was efficient and students performed better. This vertical integration was adopted because nanotechnology encompasses, yet is different, for all curricula, and thus cannot be easily compartmentalized and taught like a traditional subject. We also observed that hands-on learning greatly facilitated students grasp the fundamentals of nanoscience and nanotechnology.

Introduction

Nanotechnology is a popular catchphrase, however, the definition is unclear to students. We define Nanotechnology as "the fundamental understanding of distinctive phenomena at the nanoscale, 1-100 nanometers in size, and to develop structures, devices and systems that possess novel properties and functions because of their unique size." What are these special properties? They are the physical, mechanical and electrical properties of matter that are enhanced at the nanoscale due to the small size and perfect construction of the material. Some of these properties depend on the geometry of the material, not just the chemical composition. For example, the electrical properties of carbon nanotubes depend on the twist orientation of the atoms and the diameter of the tube. Thus, broadly defined, nanoscale science and technology may be the fastest growing and most diverse science at present in the world. This is an emerging science and multiple disciplines are involved in this field and there is not much background or infrastructure available to teach a course like Nanotechnology to a wide audience. Furthermore, since this is not a traditional course, faculty do not have experience teaching nanotechnology. At the University of Cincinnati, we have developed a vertical integration concept to integrate nanoscale science and technology into the curriculum. These include:

- A Plan for Vertical Integration of Nanotechnology

- Introducing Nanotechnology to Middle School Students

- Nanotechnology Undergraduate Education (NUE)

- Undergraduate Research in Nanoengineering

- Nanotechnology Graduate Research and Education

Our experiences are described in the following sections. We expect that these experiences will provide a model that other faculty could potentially use to design similar courses in their institutions.

Plan for Vertical Integration of Nanotechnology

The plan is to integrate nanotechnology into the academic community from middle school, to high school, to community college and graduate school and beyond. Our approach is based on vertical integration of teaching, based on related work by Pai and coworkers, where computer software was integrated into an engineering curriculum(*1*). Briefly, vertical integration is a staged learning process in which students are broadly trained in some transforming technology from the bottom up to make learning easier and more interesting. For example, computer software (e.g. MATLAB, MS Office) can be taught in graded stages and used continuously in different courses from grade school to graduate school. Nanotechnology can follow a similar path. There can also be parallelism between teaching and research at the higher levels. A similar vertically integrated nanotechnology teaching and research plan used in engineering at UC (*2*) is outlined in the flow chart in Figure 1.

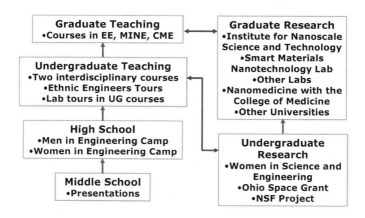

Figure 1. Vertical integration of nanotechnology education at UC.

To continuously improve the capability of the teachers in the vertical teaching process, a back propagation method is proposed to continuously train teachers at each level to keep up with rapid changes in the field. This is important because the number of innovations and publications in this area is increasing rapidly. In the scheme shown in Figure 2, students trained at each level would train students at their previous lower level to flow the new information backward through the curriculum within a short time span. A back propagation neural network works in a similar manner. This approach to teaching nanotechnology employing significant teaching by students is interactive, provides feedback, and allows students to mentor junior students. This steady approach to learning is expected to increase acceptance and understanding of nanotechnology as opposed to introducing a new subject at a higher academic level. Teaching in a top down approach is also an efficient use of faculty resources and students tend to be more open to learning a new subject when it includes experiences from their upper level peers. As this approach matures, we expect that the dissemination of knowledge will become seamless at all levels

Some of the teaching activities involved in the vertical integration scheme are described below.

Introducing Nanotechnology to Middle and High School Students

UC has visitation days in which high school and middle school science teachers learn recent developments in their fields, including nanotechnology. Teachers attend seminars, visit research labs, and are introduced to novel concepts in nanotechnology. Several programs are also underway to teach engineering and nanotechnology to high school and middle school students. These programs are Men In Engineering Summer Camp; Women In Engineering

Summer Camp; and Special Seminars at Middle Schools. An overview of some of these programs is described below.

Figure 2. Flow-down training of teachers.

A. *Nanotechnology Seminars for Middle School Students*

This hands-on program is for 7th and 8th grade students from several Cincinnati Public Schools. The objective of the event was also to increase participation of underrepresented students in mathematics, science, and engineering. Briefly, flow-down teaching was used in which an undergraduate or graduate student would present a seminar on nanotechnology to the middle school students(*3*). Next, students were asked to build C_{60} bucky balls using chemistry sets. (Figure 3).

Figure 3. Middle school students building a carbon buckyball.

The hands-on experience of making nano-structures helped students understand how geometric structure influences physical properties. Many students expressed interest in the field and were intrigued by the applications of nanotechnology. A faculty member and senior graduate students also participated in this program.

B. Nanotechnology Seminars for High School Students

This program is similar to the program for middle school students and follows the same format. The program is combined with the established Women in Engineering and Men in Engineering programs. The main objective of these programs is to increase US student participation in mathematics, science, and engineering and enrollment at the university. Typically, a graduate student and a faculty member give a presentation on nanotechnology as part of the UC Women in Engineering summer program for high school students (Figure 4). After the lecture to introduce nanotechnology, graduate students gave nanotechnology demonstrations including growing nanotubes and processing nanocomposite materials. The one difference between the middle and high school programs is that material covered in the high school program is a little more advanced than the middle school program.

Figure 4. Undergraduate student teaching nanotechnology to high school students in the women in engineering program.

Our experience teaching high school and middle school students indicates that nanotechnology is best taught on an introductory level. Teaching basic nanotechnology can reinforce ideas of structure versus function of materials and give students more practice with measurement scales and chemical composition. We also observed that most of the middle school students were very highly motivated and willing to learn, despite formal training in the sciences. They quickly began to understand atomic structure and molecules after seeing and putting together the nanotube, buckyball, and diamond stick models. They also became familiar with how small nano is. For middle school students, this is their first exposure to the fundamentals of nanoscience. For high school students, we introduce nanotechnology at a slightly higher level.

This program exposes students to knowledge that are not necessarily available in high school. This outreach program promotes a desire to stay in the sciences. We also observed that middle and high school students relate favorably to upper-level peers.

Nanotechnology Undergraduate Education

At UC, two new undergraduate courses, *Introduction to Nanoscale Science and Technology* and *Experimental Nanoscale Science and Technology* (²) were introduced in 2006 and have become permanent additions to the undergraduate curriculum. The courses were developed by faculty members from the Engineering, Physics, Chemistry, and Philosophy departments and was funded by the NSF Nanotechnology Undergraduate Education (NUE) program and UC. The courses were developed to incorporate nanotechnology education into undergraduate curricula(*4, 5*). The two courses are interdisciplinary and can be taken as an undergraduate elective course for students in the sciences and engineering.

We had 24 students with various backgrounds in the first year of the *Introduction to Nanoscale Science and Technology* course. The diverse topics and perspectives coupled with the backgrounds of various students required careful consideration to maintain flow and structure. We covered topics such as Background to Nanotechnology (what's hot and cool about nanotechnology, periodic table, atoms, molecules, and polymer, matter at the nanometer scale), Nanobiotechnology (Nature's nanofactory – the cell, DNA / RNA / proteins / lipids / carbohydrates, diagnostics and drug delivery). Nanoparticles (nanotubes, nanowires, nanocrystals, nanoshells/spheres), Nanomechanics (nanoMEMS, nanorobots), Nanoelectronics (semiconductors, from classical to quantum physics, quantum devices) and Nanophotonics (quantum wells, wires, dots, rings, nanolithography). A faculty member from the philosophy department covered the philosophy aspects (historical and humanistic perspectives, comparison of technical / scientific revolutions in history, societal implications and ethics). The varying perspectives of faculty who participated in the course allowed the students to view nanotechnology from different points of view. All the lectures had material and examples of functionalized nanomaterials relevant to new and exciting opportunities for research. Based on the course evaluations and in class student presentations, we believe that this course was well received. Our future goals include expanding the size of the class to a wider student body. Individual faculty also intend to introduce upper level undergraduate courses / graduate courses in the near future. These efforts have been described in more detail elsewhere(*4*).

The course *Experimental Nanoscale Science and Technology* was introduced in the spring quarter of 2006 at UC. The course covers experimental synthesis and characterization of nanostructures and the devices that can be made from them. The course was organized to cover a full cycle of experimental nanotechnology from materials synthesis to device fabrication. The experimental techniques learned in the course can be applied to industrial problems and used as a foundation for performing research in the area of nanoscale science and technology. We emphasized safety at the outset as toxicity of synthetic nanomaterials is still being researched.

The course introduces students to experimental nanotechnology through four hands-on modules. A presentation was given at the beginning of each module to familiarize students with the fundamentals of their experiments. Module 1 is "Introduction to Synthesis of Carbon Nanotubes (CNT)." This module covers Chemical Vapor Deposition (CVD) and related synthesis hardware, the catalytic mechanism, oriented growth, substrate preparation, purification of CNT, characterization using Scanning and Transmission Electron Microscopy (SEM/TEM) and Atomic Force Microscopy (AFM). Module 2 is "Nanotube Device Fabrication & Evaluation." This module uses the nanotube arrays grown in the first module to make sensor devices. The module covers transduction and electrochemical properties of nanotubes, nanotube biosensor fabrication and electrochemical characterization. Thus, the first two modules were well integrated and were taught in the engineering research laboratories(5). Module 3 is "Synthesis and Characterization of Gold Nanospheres and Nanorods." Students used wet chemical synthesis to fabricate gold nanoparticles, and characterized them using laser excitation and emission spectroscopy and AFM/TEM techniques. The optical properties of the nanomaterials were correlated with the size and shape of the nanomaterials. Module 4 is "Semiconductor Nanocrystal Quantum Dots." This module teaches the concept of quantum confinement and deBroglie wavelength, semiconductor nanocrystals, CdSe / ZnS core-shell nanocrystals, AFM characterization, laser excitation, and emission characteristics. The last two modules were taught in Chemistry and Physics laboratories.

Safety

We have included safety procedures as a separate section because this is the first time that most students have worked in an advanced laboratory. Also, the particles are very small and the toxic nature of these particles are not well studied. We (faculty and graduate students) spent considerable time explaining the safety protocols to the students. Safety lectures were given in the course before starting the laboratory modules. The first module involves growth of CNT in a sealed reactor. We ensured that all students learned about the external sensors, exhaust system and automatic shut-off valves that are activated in case of a leak. Next, we ensured that all students wore particle masks, gloves, goggles, and individual laboratory coats to ensure they are not exposed to nanomaterials. The third laboratory module involved wet chemical synthesis of gold nanoscale particles. Again, students were given appropriate instructions on handling of solvents and chemicals. Laboratory goggles and gloves were worn at all times and all reactions were performed in fume hoods. As in the case of the previous modules, care was taken to minimize the amount of nanoparticle exposure. The faculty and graduate students tested the experimental setup and design before introducing them to students. The fourth laboratory module was the characterization of CdSe/ZnS core-shell nanocrystals using AFM, and studying laser excitation and emission characteristics. Commercial quantum dots were handled under supervision of the faculty and graduate students. Standard operating procedures were followed before the lasers were turned on. Students understood and appreciated that safety is a critical component when it comes to exporing new areas of science and developing new experiments.

Overall, we observed the theme of vertical integration was translated well as undergraduate students relate favorably to graduate students and follow directions when it comes to developing nanoparticles or using advanced instrumentation.

Module 1: CVD Synthesis of Carbon Nanotube Arrays

In this module, students gained hands-on experience fabricating carbon nanotube arrays. The module is important because students are exposed to the techniques used in developing nanoarrays. A brief description of the module follows.

First, the substrate is synthesized. Appropriate multi-layered substrates ($Si/SiO_2/Fe$) for CVD growth of CNT arrays was made by senior graduate students under the supervision of experienced faculty members. After the substrates were ready, the graduate student (teaching assistant) helped the undergraduates grow the nanotube arrays using CVD. CVD is the irreversible deposition of a solid from a gas or a mixture of gases through a heterogeneous chemical reaction. A schematic of the reactor used for running the CVD process is shown in Figure 5. The heterogeneous chemical reaction is driven by heat. The formation of the solid product from gaseous reactants during the CVD process is based on either decomposition or combination type chemical reactions illustrated below.

$$A(g) \longrightarrow B(s) + C(g)$$

$$A(g) + B(g) \longrightarrow C(s) + D(g)$$

The major variables in CVD of carbon nanotubes include catalyst type, substrate design, carbon precursor, and deposition parameters such as temperature, flow rates, time, and pressure. The CVD process is generally diffusion controlled. Depending on the substrate design, a growth time of 1 hour yields a CNT array about 1 mm long.

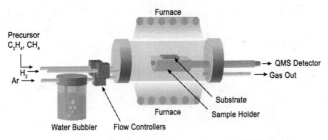

Figure 5. Schematic of a CVD Reactor.

An EasyTube Nanofurnace (6) is used to synthesize CNT's. The furnace utilizes a thermally driven CVD process and consists of two main units: the control unit, and the process unit. The process unit is comprised of a furnace and a quartz tube reactor with a loader. The control unit contains the workstation for the operator and its display. A computer controller regulates the furnace temperature and the gas flow. The system uses four process gases: methane, ethylene, hydrogen, and argon for the synthesis. Nitrogen is employed for operating the pneumatic gas valves in the furnace. The hydrocarbons methane and ethylene serve as carbon precursor gases. The use of these gases depends on the selected recipe for growing CNTs and provides a different number of carbon atoms per molecule. Hydrogen is used for controlling the volume concentration of the carbon atoms produced in the gas phase during the hydrocarbon decomposition. Argon creates an inert atmosphere and is used for the purging and cooling process steps. The CVD system is enclosed in a large fume hood with safety glass doors. In addition, leak detectors for hydrogen and hydrocarbons are installed in the gas cabinet and all along the gas supply lines to the furnace.

Under the supervision of graduate students, undergraduates grew CNT's. Several experiments were done by varying the flow rates for the gases. Students learned that there should be only one independent variable (i.e. process condition that is varied) for each experiment. The students found that long CNT's could be produced using a particular combination of variables. Students also characterized the nanotubes using atomic force microscopy (AFM). This was their first exposure to an expensive instrument. We observed that they were nervous using the AFM, but quickly became adept at the measurement of the samples. Overall, the students learned the AFM is an important characterization tool with a specific range of resolution. They also used Environmental Scanning Electron Microscope (ESEM) to characterize the nanotubes. (Figure 6)

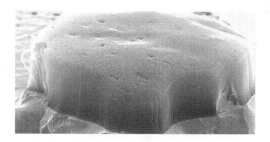

Figure 6. ESEM (Philips XL-30 Field Emission instrument) image of a carbon nanotube array grown by students. Individual CNT are about 20 nm in diameter and 1 mm long.

Module 2: Carbon Nanotube Tower Electrodes

The millimeter tall multi walled carbon nanotube (MWCNT) arrays grown in the previous module were used for device fabrication. The main purpose of this module is to develop a nanoelectrode array including characterization of the electrode fabricated from the CNT array (9-12). The MWCNT tower is easily peeled off from the Si substrate and epoxy is cast into an array. Both ends of the electrode are polished with one end for electrical connection and the other end for the nano electro-array sensor.

A detailed procedure for sensor fabrication was included in the laboratory manual for the course. Polishing the array was the trickiest step. Students had to be patient, to keep checking the electrode under the microscope and to make conductivity measurements until the nanotubes were exposed in the epoxy. The electrodes were tested using cyclic voltammetry (CV) to evaluate the electrochemical properties of the CNT electrode. CV is a linear potential waveform; the potential is changed as a linear function of time. The rate of change of potential with time is referred to as the scan rate. Measurements were carried out using a Bioanalytical Systems Analyzer. A platinum wire and an Ag|AgCl wire were used as the auxiliary and reference electrodes, respectively. All reagents were freshly prepared in deionized water by groups of two students. The CV for the reduction of 6.0 mM $K_3Fe(CN)_6$ (in 1.0 M KNO_3 as a supporting electrolyte) at a nanotube tower electrode obtained at different scan rates was measured. The CV of the sensor (Figure 7) indicated that the mass transport was dominated by radial diffusion.

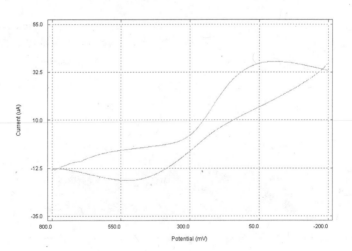

Figure 7. Results of CV of 6 mM $K_3Fe(CN)_6$ in 1.0 M KNO_3 using the CNT tower electrode with a scan rate of 100 mV/s.

In this module, students learnt to fabricate a sensor, the fundamentals of electrochemistry and measured the CV's of the CNT arrays that they had developed in the previous module.

Module 3: Synthesis of a Gold Nanorods

Nanoparticles show properties that are often different from those of their corresponding bulk materials(13). Metal nanoparticles ranging from noble to transition elements have shown interesting properties in catalysis, optics, magnetism, sensors, etc. In the synthesis of metal nanoparticles, control over the shape and size has been one of the important and challenging tasks. The size can influence the optical properties of metal nanoparticles. This is especially important when the particles have large aspect ratios (length/width). Therefore, we devised this module to expose students to wet chemical synthesis and possible ways of controlling growth of nanoparticles.

Students used a seed mediated growth method for the synthesis of gold nanoparticles and nanorods(*14*). This method comprises two steps. The first step is to synthesize gold nanoparticles of uniform size (~ 4 nm). Briefly, a $HAuCl_4$ (III) gold salt was reduced in the presence of a sodium borohydride, a strong reducing agent, to yield spherical seed particles. This reaction can be performed in water/air and at ambient temperature. Students were able to synthesize the seed solution without any problems. Next, a growth solution, comprised of the gold salt solution, a weak reducing agent such as ascorbic acid, a surfactant (Cetyl triethylammonium bromide, CTAB) and defined amounts of silver nitrate was made. By itself, this growth solution cannot yield nanoparticles, however, addition of the seed solution allows the gold salt to reduce on the "seed" and produce larger nanoparticles. The silver nitrate and the surfactant dictate the shape of the nonmaterial; appropriate amounts of silver nitrate are required to synthesize the gold nanorods of defined aspect ratios as opposed to making larger nanospheres. Thus, by adjusting the silver content of the growth solution, students were able to grow nanorods with controlled aspect ratios.

Next students characterized the nanomaterials using AFM, TEM and UV-VIS spectroscopy. TEM images of a set of nanorods are shown in Figure 8. The UV-VIS spectra is shown in Figure 9.

In this module, students gained hands-on experience with wet chemical synthesis, were given AFM and TEM training for characterization of their materials. They understood the affect of size and shape on the optical properties of the nanomaterials.

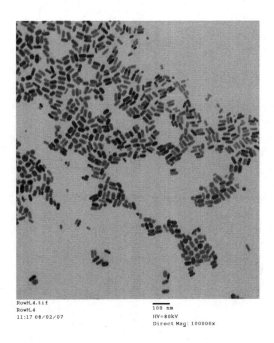

Figure 8. TEM of gold nanorods.

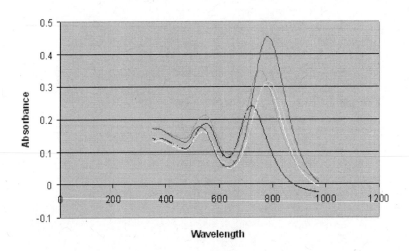

Figure 9. Gold Nanorods synthesized with varying amounts of seed and silver nitrate solution. The rods show two peaks, one near 530 nm and one between 600 - 1600 nm. Two surface plasmon bands were observed for the nanorods: the longitudinal plasmon band, corresponding to light absorption and scattering along the particle long axis (second peak); and the transverse plasmon band, corresponding to light absorption and scattering along the particle short axis (first peak).
(See page 1 of color inserts.)

Module 4: Semiconductor Nanocrystal Quantum Dots

The fourth module is the study and characterization of semiconductor nanocrystals. The objective of this module is to correlate size to optical properties and to introduce the concept of quantum confinement. Semiconductor nanocrystals allowed us to introuduce these simple concepts. Six different colloidal solutions of CdSe/ZnS core-shell nanocrystals emitting between 490 and 620 nm (1.9 to 5.2 nm radii) were purchased from commercial vendors. For absorption and emission measurements, strongly diluted solutions of nanocrystals in toluene were produced and put into standard optical cuvettes. The absorption measurements were made using a Newport MS260i spectrometer. White light after transmission through the cuvette was dispersed by the spectrometer and detected by a CCD camera. A typical set of absorption spectra for these CdSe/ZnS nanocrystals is shown in Figure 10. Students observed the blueshift in the fundamental absorption edge as the quantum dot radius reduces from 5.2 to 2.1 nm and could correlate size to optical properties. Students also used Mathematica-based modeling for realistic electrons confined to infinite one-dimensional potentials using direct solutions of the Schroedinger equation. Their analysis showed that confinement energy should increase as $1/a^2$, where a is the radius of the nanocrystal, which is the expected result.

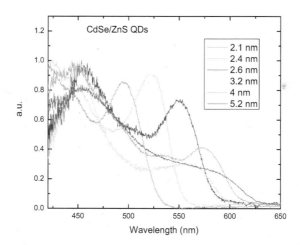

Figure 10. Absorption spectra of CdSe/ZnS core-shell QD's with diameters ranging from 1.9 to 5.2 nm. Spectra are normalized. A blueshift in the fundamental absorption edge occurs as the quantum dot diameters reduce from 5.2 to 2.1 nm.
(See page 2 of color inserts.)

Overall, we observed the theme of vertical integration was translated well as undergraduate students relate favorably to graduate students and follow directions when it comes to developing nanoparticles or using advanced instrumentation. The graduate students, under the supervision of the faculty members, prepared notes and laboratory manuals for the undergraduate students. They helped the undergraduate perform the experiments and obtain the results. The top down approach of using graduate students to teach undergraduate students worked extremely well in our experience.

Nanotechnology Graduate Research

At UC, several laboratories are actively involved in Nanoscience and Nanotechnology research. The Institute for Nanoscale Science and Technology (http://www.eng.uc.edu/ucnanoinstitute/index.php3) was established to bring researchers involved in nanoscience research together in an effort to offer promote cross cutting interdisciplinary research. Researchers involved in this effort are from Physics, Chemistry, Materials Science, Electronic and Electrical Engineering and the College of Medicine. Many nanotechnology activities are taking place at the University of Cincinnati and graduate students are at the heart of all research activities. A case in point is the Smart Materials Nanotechnology Laboratory (7) which is led by faculty from two different departments, (Professor Schulz from Mechanical Engineering and Professor Shanov from Material Science) engaged in nanotechnology research. The laboratory has three components; a synthesis, materials processing and device fabrication laboratories. This provides the capability for the full cycle of development of materials and devices based on nanoscale materials. Work in the smart nano laboratory is highly interdisciplinary and collaborative involving the colleges of Arts and Sciences, Engineering, and Medicine at UC, and other universities including North Carolina A&T State, the Air Force Institute of Technology, and the California-Berkeley. The laboratory provides exciting research and discovery for doctoral students, post-doctoral researchers, and visiting scholars.

Several students have integrated biomimetics and micro-nanotechnology to develop a structural neural system for structural health monitoring. Biomimetics provided the highly distributed parallel processing architecture needed to simplify data acquisition on large structures. Micro and nanotechnology have provided the integrated multi-state sensing capability needed to detect different types of damage, and also to monitor structural operating conditions such as temperature, strain, vibration, and pressure. Continuous sensors for the neural system are built in the laboratory using different exotic materials such as piezoelectric ceramics, carbon nanotubes, and metal nanowires. Other smart systems developed in the laboratory include an impedance biosensor developed by students in the laboratory(7).

Thus, faculty are training graduate students to become independent, interdisciplinary scientists, who will be able to solve challenging problems and interact with scientists from different disciplines to advance scientific knowledge in the field of nanoscience and nanotechnology.

Conclusions and Future Outlook

Middle school students through graduate students were taught nanotechnology within a vertically integrated framework. Middle and high school students became excited about nanotechnology. Undergraduate students were able to perform interdisciplinary nanotechnology experiments and use advanced instrumentation. The students obtained experience in synthesizing nanoscale materials, characterizing the materials, and building a nanoscale sensor device. The theoretical-experimental course sequence is considered unique and valuable because this experience would be difficult to obtain by other means such as taking different courses or through work in industry. The graduate students were capable of performing innovative nanotechnology experiments.

Initial assessments at different stages suggest that vertical integration of education and top-down teaching were efficient for teaching and provided interaction at different levels. We expect that these experiences will provide a model that other faculty could potentially use to design similar courses in their institutions.

Acknowledgements

We would like to acknowledge the financial support of the University of Cincinnati and the National Science Foundation through the Nanoscale Undergraduate Education program, grant 0532495; and the support of Newport, Inc and Nanoscience, Inc. for development of these laboratories. Students who participated in the nanotechnology laboratory and contributed to the paper are: Jonathan Abel, Charles Barbour II, Noel D'Souza, Christopher Direnzi, Donald Dlesk Jr, Zachary Kier, Nathaniel Negassi, Ryan Schaub, Kyle Seger, and Raymond Wagner.

References

1. Pai, D.; Kelkar, A.; Layton, R.; Schulz, M.; Dunn, D.; Owusu-Ofori, S. "Vertical Integration of the Undergraduate Learning Experience," **1998** ASEE Computers in Education Conference.
2. www.eng.uc.edu/nanocourses/
3. Teacher information about n the middle school program can be obtained from: Katrina Henderson; Program Coordinator for the SED program; Schwab School--Cincinnati Public Schools; 513-825-7634, email: henderka@uc.edu.

4. Bickle, J.; Iyer, S. S.; Mantei, T.; Papautsky, I.; Schulz, M.; Shanov, V.; Smith, L.; Steckl, A. Integration of Nanoscale Science and Technology into Undergraduate Curricula. *IEEE Nano 06 Conference Proceedings*, Cincinnati, Ohio, July 17-20, **2006**, 403-405.
5. Shanov, V. N. ; Yeo-Heung, Y.; Smith, L.; Iyer, S. S.; Jadhav, S.; Hoang, T. B.; Gorton, A.; Schulz, M. J.; Mantei, T.; Abel, J.; Barbour C. II; D'Souza, n.; Direnzi, C.; Dlesk D. Jr.; Kier, Z.; Negassi, N.; Schaub, R.; Seger, K.; Wagner, R.; Witt, E.; Bickle, J.; Paputsky, I.; Gerner F. First Experiences Teaching Experimental Nanoscale Science and Technology to Undergraduate Students. *IEEE Nano 06 Conference Proceedings,* July 17-20, **2006**, Cincinnati, Ohio.
6. http://www.firstnano.com
7. Smart Materials Nanotechnology Laboratory, www.min.uc.edu/~mschulz/smartlab/smartlab.html.
8. Shanov, V. N.; Yun, Y-H.; Tu, Y.; Schulz, M. J. Substrate and Process Interplay During Synthesis of Millimeter Long Multi-Wall Carbon Nanotube Arrays. *IEEE Nano 06 Conference Proceedings*, July 17-20, **2006**, Cincinnati, Ohio.
9. Yun, Y-H.; Bange, A.; Shanov, V. N.; Heineman, W. R.; Halsall, H. B.; Wong, D. K. Y.; Behbehani, M.; Pixley, S.; Bhattacharya, A.; Dong, Z.; Schulz, M. J. A Nanotube Composite Microelectrode for Monitoring Dopamine Levels using Cyclic Voltammetry and Differential Pulse Voltammetry. Submitted to the *J. Nanoengineering and Nanosystems.*
10. Yun, Y-H.; Bange, A.; Heineman, W. R.; Halsall, H. B.; Shanov, V. N.; Dong, Z.; Pixley, S.; Behbehani, M.; Jazieh, A.; Tu, Y.; Wong, D. K. Y.; Bhattacharya, A.; Schulz, M. J. *Sensors and Actuators B.*, **2007**, *123*, 177-182.
11. Yun, Y-H.; Bange, A.; Shanov, V. N.; Heineman, W. R.; Halsall, H. B.; Dong, Z.; Jazieh, A.; Tu, Y.; Wong, D. K. Y.; Pixley, S.; Behbehani, M.; Schulz, M. J. Highly Sensitive Carbon Nanotube Needle Biosensors. Submitted, *Journal of Nanotechnology.*
12. Bange, A.; Halsall, H. B.; Heineman, W. R.; Yun, Y-H.; Shanov, V. N.; Schulz, M. J. Carbon Nanotube Array Immunosensor Development. Invited Paper, *IEEE Nano 06 Conference Proceedings,* July 17-20, **2006**, Cincinnati, Ohio.
13. Wilson, M.; Kannangara, K.; Smith, G.; Simmons M.; Raguse, B. *Nanotechnology, basic science and emerging technology*; Chapman & Hall/CRC, 1st edition, 2002 .
14. Murphy, C. J.; Sau, T.K.; Gole. A.M.; Orendorff, C.J.; Gao, J.; Gou, L.; Hunyadi, S.E.; Li, T. Anisotropic metal nanoparticles: synthesis, assembly, and optical applications. *J. Phy. Chem. B.*, **2005**, *109*, 13857-13870.
15. Engineering web page; http://www.eng.uc.edu/.

Chapter 5

Big Emphasis on a Small Topic: Introducing Nanoscience to Undergraduate Science Majors

Richard W. Schwenz[1*], Kimberly A. O. Pacheco[1], Courtney W. Willis[2] and Wayne E. Jones, Jr.[3]

1. School of Chemistry and Biochemistry, University of Northern Colorado, Greeley, CO 80639
2. School of Earth and Physical Sciences, University of Northern Colorado, Greeley, CO 80639
3. Department of Chemistry and Institute for Materials Research, Binghamton University (SUNY), Binghamton, NY 13902
   *. Corresponding author

> A program bringing nanoscience concepts into the science major and pre-service teacher curricula is described. This program includes learning communities, laboratory exercises, and internships.

What do casual pants, golf balls, *National Geographic* (*1*), and drug delivery systems (*2*) have in common? All are advertised as containing or giving information about nanomaterials because of the public enchantment with nano- anything, even if the nanomaterials present in the formulation are not new. In business, major investments are being made in nano –as the area is viewed as the new frontier of science. In 2007, the US government budgeted close to 1.3 billion dollars for nanotechnology research and development (*3*). This continuing investment requires citizens to be able to make informed decisions about whether this investment is financially, ethically, and scientifically meritorious. To properly educate the general population, nanoscale science concepts and experiences should be incorporated into the undergraduate curriculum, particularly for future K-12 teachers as they have the greatest influence on future citizens. Incorporation of these concepts into K-12 teacher education will ensure that nanoscale science concepts are introduced to students before they reach college through appropriate materials (*4*). Scientists need to play an active role in educating and training students in this area by developing and providing appropriate learning opportunities in the area of nanoscale science to undergraduates. Some materials are already available for nanoscale science education in print (*4, 5*) and more on the web (*6*) Previous work has developed curricula for non-science majors and elucidated some student difficulties in interpreting nanoscale phenomena (*7*). Other authors have described a course of non-science majors along similar lines (*8*).

In 2005, our science departments began developing Project *DUNES* (*Developing Undergraduate Nanoscale Experiences in the Sciences*) to enhance our undergraduate educational experiences by incorporating nanoscale science into our majors' curricula, especially at the introductory level. We made several decisions about the experiences based on the existing curricula in the sciences. First, we could not replace any of the existing introductory courses, or require any additional courses, because of the number of courses which are required for the science majors, especially our secondary science teaching licensure majors. This meant that we needed to develop teaching modules for existing courses rather than new courses. Second, we wanted to emphasize the interdisciplinary nature of work in nanoscience by using a common piece of equipment in courses across multiple disciplines. Third, we wanted to have students help develop the experiences, rather than having faculty perform all the development tasks. In addition to impacting the science majors, the project has allowed incorporation of nanoscience into the science courses taken by pre-service elementary teachers.

Our project includes three major components. The first component is a learning community for first-year science majors, especially secondary teaching majors in the sciences and mathematics. The second component is a group of laboratory instructional modules developed for courses across biology, chemistry, and physics using a common piece of nanoscience instrumentation. The modules were an attempt to stress recent research in nanoscience rather than simply repeat other previously developed materials. The third component includes internships for upper level students (with a preference given to pre-service teachers) involving module development.

Learning Communities

Learning Communities (LC) is our university's term for small groups of students of similar interests, who arrive at campus simultaneously, are placed into a defined set of courses in the first semester based on their interest, and who have a closer interaction with faculty in their interest area. Students self-select into different LC at their summer orientation prior to arrival on campus for their freshman year. The intention is to provide the support that the first generation, minority students characteristic of our institution need to help them be successful in higher education. Other campuses may use the terms freshman interest group (FIG), cohorts, residential academic program (RAP) or first year experience (FYE) to describe similar programs. The model we use enrolls students in core General Education courses, major-specific courses, and an interdisciplinary ID 108 (name of general course) class. The general ID 108 syllabus allows variation in discussion topics based on a major, career choice, or interest area. LC are intended to promote collaborative work and intellectual connections. The formation of LC has been shown to increase retention and student learning by 3-6% even with students having lower entrance scores, especially when accompanied by Supplemental Instruction (SI) (*9*). SI provides students with a tutor who has taken the (typically) large lecture course successfully, who attends all course lectures, models effective student behavior, and provides three weekly one-hour tutorial sessions. Learning communities have been shown to result in statistically significant increases in GPA, student satisfaction, and retention (*10, 11*). As a result, many universities are developing similar experiences under a variety of names.

Based on the previous success of the LC model at our university, we implemented a LC for incoming freshman science majors based on nanoscale science. Our nanoscale science LC is composed of students taking their first mathematics course, their first science course in their declared major, an interdisciplinary course (ID 108, here, *Introduction to Nanoscale Science*), and a common section of English composition (ENG 122). Mathematics courses (either College Algebra, or Calculus 1) are such an important part of, and prerequisite for, future science courses it was crucial that they be included in the LC course selections. Each student also enrolls in the first course in their major (typically cellular biology, general chemistry, or introductory physics). Within this course, we try to enroll students from a LC in a common laboratory section. We chose not to require enrollment in two science courses because data suggests that concurrent enrollment in two laboratory science courses by incoming freshmen significantly decreases student retention and learning (*12*). We bring the entire community together in the team-taught ID 108 course where students are able to work in a small group environment, bringing the students' differing science experiences to bare on problems in nanoscale science. The text used for this course is *Nanotechnology: A Gentle Introduction to the Next Big Idea* (*13*).

Students enrolling in the nanoscale science LC must have declared a science major, a secondary science teaching major, or an elementary education liberal arts major with a science concentration. In Colorado, students becoming teachers are licensed to teach either at the K-6 level (with a general certification) or 7-12 level (with secondary science certification and a content area major). Our secondary pre-service teacher programs are B.S. degrees in each of the four areas. Program requirements include the first year courses in all four areas, and further course requirements in the subject area major. The K-6 program has developed as a broadly based liberal arts program with a requirement for a concentration. In science, the four concentrations correspond to biology, chemistry, earth science, and physics. We assume that this combination of students allows the LC to be relatively cohesive, brings multiple viewpoints to discussions, and includes students who will affect the lives of others.

Table 1. Week-by-week events in the learning community course

Week	Topic	Activity / Assignment
1	Library Resources	Tour of Campus Library
2	What is Nanotechnology?	Introduction to Size Scale / Pre-test
3	Continuing Discussion of Nanotechnology	Summarize an article concerning Nanotechnology
4	Tools used in Nanotechnology	LEGO™ AFM (Detecting an atom)
5	Atomic Force Microscopy	Hands – on use of AFM to view CD sample
6	Scanning Electron Microscopy	Visit the UNC Imaging Suite User Facility
7	Staying on Track	Academic Advising Quiz, Four Year Plan of Study
8	The Ultimate Road Trip	How to survive your first year at college presentation
9	Societal and Ethical Considerations regarding Nanotechnology	Medical Nanobot Societal Impact Activity (*14*)
10	Nanotechnology and Cancer	Benefits vs. Safety Risks Discussion
11	Scanning Tunneling Microscopy	Indirect Observations: Atoms and the STM
12	Scanning Tunneling Microscopy	Hands – on use of STM with graphite samples
13	Nanowires and Nanotubes: Medicine and Homeland Security	In – class discussion
14	Wrapup	Post – test

Other Activities:
    Required freshman workshops
    Introduction to campus resources / technology (computer labs)

The learning community course, *Introduction to Nanoscale Science*, brings the students together to focus on many aspects of nanoscale science. The standard LC course description has required elements relevant to a freshman's transition into the university life including library utilization, campus technology, a 4-year educational plan, and career exploration. The course elements are chosen to help students 1) gain an understanding of the goals, culture, and values of higher education; 2) connect with faculty, peers and advisors; 3) practice critical thinking, problem solving, written and oral communication skills; 4) use resources for course-related projects; and 5) explore academic areas. On top of these elements, we add the goal of students' gaining an appreciation for the field of nanoscience and its societal implications. Table 1 shows a period by period listing of the ID 108 class activities during each class meeting. Assignments have students reading current literature articles related to nanotechnology at both the scientific and lay person level. Students involved in this LC develop a heightened appreciation for developing technology and the interdisciplinary nature of nanoscale science from the beginning of their college career as a result of this experience.

General Science Laboratory Experiences

The second component of our project is the incorporation of nanoscale science experimentation into the introductory level curricula in each science discipline. In doing so, we emphasize topics of current research interest in nanoscale science and use a common instrument across traditional disciplines. We incorporate a modular approach to facilitate replacement of existing laboratory exercises rather than the replacement of an entire course, or the development of an additional course that might be poorly populated as an elective course in a number of majors. Each module is designed around the use of a scanning probe microscope (SPM), operated as either a scanning tunneling microscope (STM) or as an atomic force microscope (AFM), independent of the scientific discipline involved. We incorporate the SPM because it is commonly used in nanoscale science experiments and because one way to illustrate the field's interdisciplinary nature is to demonstrate the multiple ways one instrument can be used in several traditional fields to obtain results of interest to that field. Thus each module makes use of a SPM and a model for a SPM to instruct the students in a specific science discipline, and simultaneously in the fundamental concepts of nanoscale science. The modules themselves can be found online (15) and will be addressed more completely elsewhere.

Two modules are being incorporated into the chemistry major curriculum. These include a module on self-assembled monolayers (SAM) and one on the synthesis and characterization of quantum dots. The SAM experiment is designed for a second semester general chemistry course. It may be used as an intermolecular forces lab experiment or as an introduction to materials science / nanoscience. The laboratory experiment exposes students to the chemistry of silanes and self-assembled monolayers. Students gain hands-on experience with a contact angle instrument and an AFM over two laboratory sessions while learning how and why self-assembly occurs. During the first session, students prepare their samples for imaging and the second session allows for measuring contact angles and imaging the samples.

The quantum dot module, intended for the undergraduate physical chemistry laboratory, outlines the preparation of CdSe quantum dots using techniques given elsewhere (16, 17). The absorption and fluorescence spectra are collected under a range of preparation conditions that vary the size of the dots. The larger dot results in a longer wavelength absorption feature. The spectra are compared with that predicted by a simple model.

The module developed to incorporate biological concepts of nanoscale science involves the regeneration of damaged snail shells (18). Shells can be damaged in a number of ways, including predator attack, wave action, or disease. Shells are formed from thin layers of calcium carbonate and protein. Although the specifics of the shell formation and regeneration are unknown, it appears that regeneration involves a self-assembly of calcium carbonate, possibly under the direction of shell proteins. The regrowth rate is typically 6 microns per day. One of the reasons this nanoscale regrowth is interesting because understanding it may lead to devising ways to regenerate bone tissue. During this laboratory experience the students damage the shell of live marine snails in order to observe the regrowth process. Students gain an appreciation for the natural process of self-assembly involved in the shell regeneration, learn how nanotechnology may be able to mimic the regeneration process, and experimentally observe the regrowth over several weeks.

Two modules have been incorporated into the elementary pre-service teachers' required physical science course. The first is a modification of an earlier exercise from the University of Wisconsin (*19*). This experience focuses on teaching introductory students about surface analysis techniques by constructing a macroscopic model using the same principles in constructing a nanoscopic AFM scan head We extended these materials to allow greater student creativity through a wider collection of LEGO^{TM} bricks, through a hands-on experience modeling how the AFM works, and then using a real AFM instrument to image the holes in the plastic of a compact disk. The students learn how an AFM works and can apply that knowledge to better understanding the image obtained by the actual instrument.

The second module the pre-service teachers experience uses a scanning tunneling microscope to image highly-ordered pyrolitic graphite (HOPG) (*20*). This module focuses on familiarizing students with STM. They are encouraged to view earlier scans and to collect scans and images. Worksheets are offered to aid in observing if the students understand what they are doing. The worksheet and questions are instructor extendable. The module is broken into 6 sections so other instructors can use them as desired. For more advanced students, questions can be asked on concepts such as electron tunneling, logarithmic functions, and electrical current. This activity is meant to be completed with HOPG, but can be done easily with any standard sample where the outcome is clear to the students. At the completion of the modules, the students are expected to have an understanding of how an image of HOPG is obtained using the STM and what samples can be used for imaging with STM.

Conclusion

Project *DUNES* includes several goals related to increasing student learning of nanoscale science concepts during their university education. Nanoscale science concepts are incorporated into a variety of experiences extending throughout the university experience. These concepts are addressed by formal instruction during lecture and laboratory experiences across the curriculum from the introductory to advanced level, and by informal, structured experiences allowing students to bring nanoscale science experiences into their life after college. Specifically, the lower division (first and second year) curricula for future science majors, secondary science education majors, and some elementary education majors are enhanced by the development of a learning community focused on nanoscience and by incorporating nanoscale science modules into their existing course structure. By itself, performing these new experiences builds some nanoscale science competence. However, simply incorporating some modules deemphasizes the interdisciplinary nature of nanoscale science and the cooperation between disciplines that characterizes the science as it is currently performed. In this project, the interdisciplinary and cooperative nature of science is emphasized and strengthened by building a learning community focused on nanoscale science from the first year, and by using the same instrument in multiple science classes to explore aspects of nanoscale science problems from the viewpoints of different sciences. We were also able to enhance the learning of selected future teachers by awarding summer internships for materials development. We envision the development of student learning emphasizing the cooperation between disciplinary individuals, their different approaches to problems, and the idea that cooperative research is a better approach to solving difficult problems.

Acknowledgements

The authors would like to acknowledge the National Science Foundation for award DUE-0532516 which funded this project.

References

1. Kahn, J. *National Geographic* **2006**(6), 98-119.
2. Allen, T.M. and Cullis, P.R. *Science* **2004**, *303*, 1818-1822.
3. 2007 Federal Budget Fact Sheet. http://www.ostp.gov/html/budget/2007/2007FactSheet.pdf (accessed October, 2007).
4. Jones, M.G., Falvo, M.R., Taylor, A.R. and Broadwell, B.P. *Nanoscale Science*; National Science Teachers Association: Arlington, VA, 2007.
5. Sweeney, A.E. and Seal, S., Eds. *Nanoscale Science and Engineering Education*; American Scientific Publishers: Stevenson Ranch, CA, 2008.
6. NanoEd Resource Portal. http://www.nanoed.org/ (accessed April, 2008).
7. Tretter, T.R., Jones, M.G., Andre, T., Negishi, A. and Minogue, J. *J. Res. Sci. Teach.* **2006**, *43*, 282-319.
8. Crouch, R.D. *J. Coll. Sci. Teach.* **2006**, *36*,(1), 40-44.
9. Webster, T.J. and Hooper, L. *J. Chem. Educ.* **1998**, *75*, 328-331.
10. Zhao, C.-M. and Kuh, G.D. *Res. in Higher Educ.* **2004**, *45*, 115-138.
11. Hotchkiss, J.L., Moore, R.E. and Pitts, M.M. *Educ. Econ.* **2006**, *14*, 197-210.
12. Seymour, E. and Hewitt, N.M. *Talking about Leaving: Why Undergraduates Leave the Sciences*; Westview Press: Boulder, CO, 1997.

13. Ratner, M.A. and Ratner, D. *Nanotechnology: A Gentle Introduction to the Next Big Idea*; Prentice Hall: Upper Saddle River, NJ, 2003.
14. Societal Implications Activity: Rocks and Nanobots. http://mrsec.wisc.edu/Edetc/IPSE/educators/socImp1.html (accessed April, 2008).
15. NUE project Web site. http://mast.unco.edu/nue (accessed April, 2008).
16. Boatman, E.M., Lisensky, G.C. and Nordell, K.J. *J. Chem. Educ.* **2005**, 1697-1699.
17. Garland, C.W., Nibler, J.W. and Shoemaker, D.P. Spectroscopic Properties of CdSe Nanocrystals. In *Experiments in Physical Chemistry*; McGraw-Hill: Boston, MA, 2009; p. 492.
18. Morrow-Baker, A.C. and LaCrue, S. *Current: J. Marine Educ.* **2006**, *22*, 18-22.
19. Exploring the Nanoworld with LEGO® Bricks. http://mrsec.wisc.edu/Edetc/LEGO/bookindex.html (accessed April, 2008).
20. Zhong, C.-J., Han, L., Maye, M.M., Luo, J., Kariuki, N.N. and Jones, W.E. *J. Chem. Educ.* **2003**, *80*, 194-197.

Nanoscale Laboratory Experiences

Chapter 6

Introduction to Powder Diffraction and its Application to Nanoscale and Heterogeneous Materials

Robin T. Macaluso

School of Chemistry and Biochemistry, University of Northern Colorado, Greeley, CO 80639

> Diffraction is a useful tool to learning the structure of a crystalline material. The first part of this chapter highlights basic crystallography topics, powder diffraction theories and analysis methods. Diffraction is also effective in studying the structure of nanoparticles and heterogeneous nanostructures; however, these materials require instrumental and analytical adaptations. Hence, peak broadening and small-angle X-ray scattering are briefly discussed in the second part of this chapter.

Introduction

Diffraction techniques have been successfully applied and modified to learn about the structure of a wide variety of materials including magnetic materials, superconductors, glasses, micelles, proteins, polymers, and inorganic catalysts. Samples can be single crystal or powder, solid or liquid (although less commonly so). This chapter discusses the underlying diffraction principles, surveys common experimental and analysis techniques, and briefly discusses applications of diffraction for nanostructured materials.

Diffraction is a useful tool for studying the periodic structure of solid materials. Why solids? Most introductory physics textbooks approach the topic of diffraction in light of the progression of multiple wavefronts through a series of regularly repeating slits. Once the wave fronts pass through the openings, parts of each of the waves will interfere constructively and other parts of the two waves will interfere destructively.

Just as in the previous example where diffraction results from the regularly repeating slits, a real sample must have a repeating structure for diffraction to be observed. Gaseous samples consist of molecules weakly bound to each other; the attraction between liquid molecules is greater, but is still not sufficient to obtain structural order. Solids, on the other hand, possess the greatest degree of structural order; thus diffraction experiments generally provide insight into the structure of crystalline solids characterized by long range order. Amorphous solids, such as glass, still possess high degree of disorder although they have a fixed volume. Small angle-scattering has been developed to address such materials and is briefly dicussed later in this chapter.

Crystalline materials

A crystalline solid is one where a basic unit can be defined such that repeating the unit in three dimensions results in a complete description of the entire structure. In a crystalline material the basic unit repeated in three dimensions is called a unit cell, and it alone contains all the symmetry operations of the entire structure. Each unit cell can be described with unit cell dimensions, *a*, *b*, and *c*, and three angles, α, β, and γ.

Atoms and molecules in solids arranged in a lattice can be related by four crystallographic symmetry operations – rotation, inversion, mirror, and translation – that give rise to symmetry elements. Symmetry elements include rotation axis, inversion center, mirror plane, translation vector, improper rotation axis, screw axis, and glide plane. The reader interested in symmetry and solving crystal structures from diffraction data is encouraged to refer to other sources (*1-4*).

The crystallographic symmetry elements can be defined in three orientations, giving rise to seven crystal classes: triclinic, monoclinic, orthorhombic, tetragonal, hexagonal/trigonal, and cubic. The definitions of these crystal classes are shown in Table I. In addition to the crystal class, lattices can be centered such that not all lattice points occupy the corner positions of the unit cell. The seven centering possibilities are described in Table II. Combining

Table I. Unit Cell Dimensions in Various Crystal Classes

Crystal class	Lattice Parameters	Angles
Triclinic	$a \neq b \neq c$	$\alpha \neq \beta \neq \gamma \neq 90°$;
Monoclinic	$a \neq b \neq c$	$\alpha = \gamma = 90°$; $\beta \neq 90°$
Orthorhombic	$a \neq b \neq c$	$\alpha = \beta = \gamma = 90°$
Tetragonal	$a = b \neq c$	$\alpha = \beta = \gamma = 90°$
Trigonal	$a = b \neq c$	$\alpha = \beta = 90°$; $\gamma = 120°$ (H)
	$a = b = c$	$\alpha = \beta = \gamma$ (R)
Hexagonal	$a = b \neq c$	$\alpha = \beta = 90°$; $\gamma = 120°$
Cubic	$a = b = c$	$\alpha = \beta = \gamma = 90°$

The trigonal crystal system can be defined by hexagonal or rhombohedral axes, denoted by (H) and (R), respectively.

Table II. Seven Lattice Centering Choices

Lattice Centering	Symbol	Lattice Centering	Symbol
Primitive	P	Base-Centered	C
Base-Centered	A	Body-Centered	I
Base-Centered	B	Face-Centered	F

the seven crystal classes with the various centering possibilities gives rise to the fourteen unique Bravais lattices shown in Table II.

Within each unit cell are families of crystallographic planes that intersect at lattice points. Planes within a family are parallel to each other and are equally spaced. The interplanar spacing is defined as the distance between two parallel planes. A family of planes is labeled with three integer indices, generally known as h, k, and l or Miller indices. A plane is identified by the Miller indices in parentheses: (hkl). Each integer is inversely related to the unit cell dimension such that the unit cell dimensions, a, b, and c, are divided into h, k, and l parts, respectively. Examples are shown in Figure 1. A (101) plane, for example, intersects the a- and c-axes at $a/1$ and $c/1$ and intersects the b axis at infinity. For the crystallographic plane (020), a/∞, $b/2$, and c/∞ defines a plane that intersects the a-axis at the point, $1/\infty = 0$, the b-axis at the point, ½, and the c-axis at the point, $1/\infty = 0$.

The magnitudes of a, b, and c can be calculated, depending on the Bravais lattice type. Table III lists the appropriate equations for determination of unit cell dimensions based on the (hkl) indices.

Table III. Interplanar Spacings and *(hkl)* Indices

Crystal Class	Formula
Cubic	$\dfrac{1}{d^2} = \dfrac{h^2+k^2+l^2}{a^2}$
Tetragonal	$\dfrac{1}{d^2} = \dfrac{h^2+k^2}{a^2} + \dfrac{l^2}{c^2}$
Orthorhombic	$\dfrac{1}{d^2} = \dfrac{h^2}{a^2} + \dfrac{k^2}{b^2} + \dfrac{l^2}{c^2}$
Hexagonal	$\dfrac{1}{d^2} = \dfrac{4}{3}\left(\dfrac{h^2+hk+k^2}{a^2}\right) + \dfrac{l^2}{c^2}$
Monoclinic	$\dfrac{1}{d^2} = \dfrac{1}{\sin^2\beta}\left(\dfrac{h^2}{a^2} + \dfrac{k^2\sin^2\beta}{b^2} + \dfrac{l^2}{c^2} - \dfrac{2hl\cos\beta}{ac}\right)$
Triclinic	$\dfrac{1}{d^2} = \dfrac{1}{V^2}\left[\begin{array}{l} h^2b^2c^2\sin^2\alpha + k^2a^2c^2\sin^2\beta + l^2a^2b^2\sin^2\gamma \\ 2hkabc^2(\cos\alpha\cos\beta - \cos\gamma) + 2kla^2bc(\cos\beta\cos\gamma - \cos\alpha) \\ + 2hlab^2c(\cos\alpha\cos\gamma - \cos\beta) \end{array}\right]$

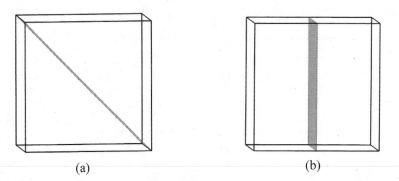

Figure 1. Crystallographic planes are shown in gray for Miller indices (hkl) of (a) (101) and (b) (020)

The Electromagnetic Spectrum: X-Rays

Electromagnetic radiation consists of transverse waves where vectors of the electric field are perpendicular to vectors of the magnetic field. The wavelength of electromagnetic radiation spans between 10^{-12} and 10^0 meters. The human eye is most effective at detecting scattered visible light – electromagnetic waves between 10^{-7} and 10^{-6} meters. X-ray radiation is defined as the portion of the electromagnetic spectrum with a wavelength between 10^{-11} and 10^{-8} meters or approximately 0.1 to 100 Ångstroms (Å). X-rays between 0.5 Å and 2.5 Å are most similar in size to atoms and bond lengths, thus we can use X-rays to probe the crystallographic structure of materials.

Bragg's Law

Bragg's Law relates the incident wavelength, λ, to the interplanar spacing, d, between (hkl) planes. Specifically, $n\lambda = 2\Delta = 2d \sin \theta$. X-rays are scattered by electrons when the conditions specified by Bragg's law are met (5). Figure 2 depicts a scenario of scattering by lattices in a material. Consider a propagating wave front of parallel beams of uniform wavelength in the X-ray portion of the electromagnetic spectrum. The X-ray beam is incident upon the crystallographic planes in a crystal, forming an angle, θ, between the (hkl) plane and the wave front. In this view, the crystallographic planes of a crystal act as mirrors so that the X-ray beams are reflected in a reflected wave front, forming a second angle, θ. Because the wave front consists of many parallel beams, the scattering event can be assumed to occur simultaneously along other (hkl) planes within one (hkl) family of planes. Two waves are reflected such that there is a path difference, labeled Δ, between the two waves. If the interplanar spacing is labeled as d, then $\Delta = d \sin \theta$. Because both k_0 and k_1 are considered in diffraction, two path differences must be summed together such that $2\Delta = 2d \sin \theta$. Constructive interference occurs when the total path difference, 2Δ, is equal to any whole number of wavelengths, $n\lambda$, where n is an integer and λ is the wavelength of the X-ray beam.

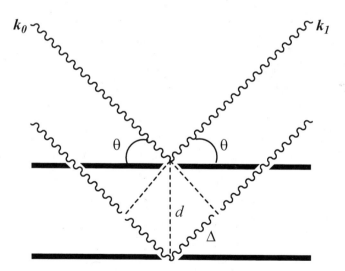

Figure 2. *The relationship between scattering angles, θ, of two wavefronts scattered by two (hkl) planes results in Bragg's Law.*

The Ewald Sphere

Another representation of Bragg's Law is elegantly displayed by the Ewald sphere (6) shown in Figure 3. A wave vector, k_o, is defined as $|k_o| = 1/\lambda$, where λ is the wavelength of the X-ray beam. The wave vector, k_o, will be scattered such that the scattered wave, k_1, will have the same magnitude as k_o. The angle between k_o and k_1 is defined as 2θ. If k_o is placed such that its tail is placed at a lattice point, then the tail of k_1 must also be placed at another lattice point such that the distance between the tails of k_o and k_1 is labeled by d^*. Thus a sphere with radius, $1/\lambda$ is defined and is commonly referred to as the Ewald sphere. The concept of the Ewald sphere reinforces Bragg's law and further illustrates that a diffraction wave is collected at an angle of 2θ, and not simply θ.

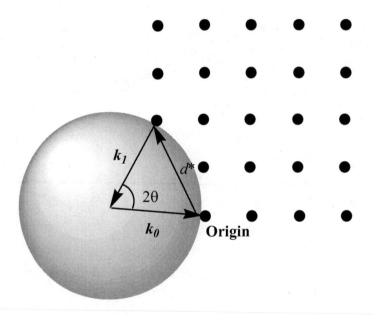

Figure 3. The Ewald Sphere

Instrumentation

The three most critical parts to an X-ray diffractometer are X-ray tube, detector, and the sample. With an appropriate generator, incident X-rays will be produced from the X-ray tube. The Ewald sphere concept dictates that the detector shall be placed at an angle 2θ from the X-ray tube.

Most X-ray diffractometers are arranged in θ-2θ or θ-θ geometry, shown in Figures 4a and b, respectively. In the θ-2θ geometry, the sample and detector arm are synchronized to form the 2θ angle. The θ-θ geometry, perhaps the most common, requires that both the X-ray tube and detector arm move in conjunction with each other. The advantage of using θ-θ geometry is that the powder sample is held in a stationary, avoiding loss of sample.

Furthermore, the Bragg-Brentano geometry utilizes focusing optics – typically focusing slits – to minimize divergence of the incident X-ray beam to maximize intensity and resolution of a flat sample. As shown in Figure c, slits are strategically placed to mimimize divergence of the incident and diffracted beams. This geometry also enables the user to easily prepare flat samples, thereby reducing spillage of powder material.

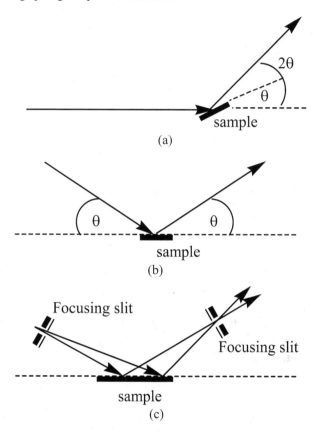

Figure 4. Powder X-ray diffraction geometries: a) θ-2θ, b) θ- θ, and c) Bragg-Brentano.

Generation of X-Rays

X-rays are generated with sealed X-ray tubes consisting of a cathode and an anode made of the target material. Typical anode materials include copper, molybdenum, etc.. When a sufficiently high voltage and current are supplied to the anode, high energy electrons strike the cathode. The resulting X-rays exit the sealed tube through beryllium windows.

The resulting incident X-ray beam will diverge; thus, a series of optics usually follows the X-ray tube. Soller slits are comprised of thin parallel plates. When placed in the beam path, the slits limit the divergence of the beam perpendicular to the plane in which the diffracted beam intensity will be measured. A crystal monochromator or β-filter can also be applied to the incident beam to suppress X-ray fluorescence.

A second divergence slit limits the in-plane divergence of the beam. The diffracted beam can also pass through a second Soller slit or monochromator before striking the detector. Scatter slits may be used to reduce background.

Detectors in diffraction are typically point detectors or area detectors. Area detectors enable faster data collection speed increases and the ability to reduce sample volume.

Diffraction Data

During typical X-ray diffraction experiments, intensity of the diffracted beam is measured as the 2θ-angle is changed; most data collection software programs will plot intensity on the y-axis and 2θ on the x-axis. Plots with d-spacing on the x-axis can also be useful and are more common with high-intensity pulsed sources that provide beams with simultaneous multiple wavelengths.

Bragg's law dictates that the value of the 2θ peak results from the interplanar spacings in in the material. It is the electron density of each atom that results in scattering of X-ray beams; therefore, the scattering factor of each element is unique. Shown below in Figure 5 are calculated X-ray diffraction patterns of NaCl and KCl (assuming Cu radiation with $\lambda = 1.5406$ Å). Both salts possess the same face-centered cubic structure so peaks with the same (hkl) indices occur for both materials; however, there are two distinct differences in the data:

- The peaks in KCl consistently shift to the left with respect to NaCl. For example, the (200) reflection occurs at approximately 32.8° for NaCl and 28.4° for KCl. Because K^+ ion is larger than Na^+, the d-spacing in KCl is larger than the analogous d-spacing in NaCl. According to Bragg's law, d and θ are inversely related, so the peaks in KCl shift to smaller θ values.
- The relative intensities between KCl and NaCl vary because of the differences in electron density ($Z = 19$) than Na ($Z = 11$). For example, the relative intensity ratio, $I_{(200)}/I_{(220)} = 1.5$ in NaCl, but $I_{(200)}/I_{(220)} = 29.2$ in KCl.

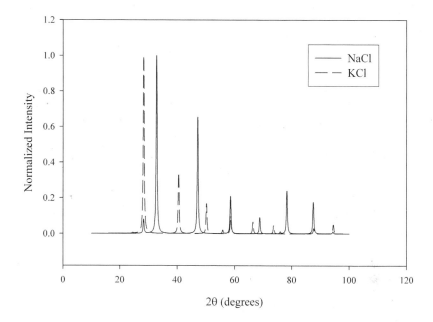

Figure 5. X-ray diffraction powder patterns of NaCl and KCl are shown as solid and dashed lines, respectively. Calculated using PowderCell(21)

Data Analysis

An important first step in analyzing diffraction data – whether it is powder or single crystal – is to index the data. Indexing involves generating a list of all possible *hkl* indices for the Bragg angles collected, determining unit cell dimensions with the *hkl* indices, assigning *hkl* indices to the observed Bragg peaks, and refining the unit cell dimensions using observed θ values. Because of the inherent high degree of overlap among reflections in powder diffraction diffraction patterns, it is typically best to have *a priori* knowledge of the crystal class or to at least have a set of selected crystal classes from which to begin indexing. Diffraction patterns can be easily compared to those in large databases, e.g., the Powder Diffraction File and The Inorganic Crystal Structure Database (*7*). *Ab initio* indexing is also possible through programs such as Crysfire (*8*).

The next step in data analysis is the determination of systematic absences to find the space group symmetry. Systematic absences are Bragg peaks that are missing in the diffraction data because of a translational symmetry operation and/or preferential orientation of polycrystalline samples.

Software package such as General Structure Analysis System (GSAS) (*1, 9*) and Rietica (*10*) are based on Rietveld methods and allow one to analyze the entire diffraction pattern at once instead of analyzing each diffraction peak individually. Essentially, the Rietveld method enables the user to create a structural model and compare the peak positions, intensities, and shape profiles in the data to those in the structural model (*11, 12*). These programs are available to download from http://www.ccp14.ac.uk.

The position of electron density (atomic coordinates) is the largest contributor to peak intensity, although temperature, preferred orientation, and polarization may also influence the peak intensities.

Peak shape is an important feature in the data. Peak shape is a convolution of three functions: instrumental broadening, wavelength dispersion, and specimen. The first two functions are dependent upon the diffractometer and its optics; it can be mathematically modeled by a combination of Gaussian, Lorentzian, Psuedo-Voigt, and Pearson-VII functions. Gaussian and Lorentzian functions describe rounded maxima and tails near the base of a peak, respectively. Most peak broadening can be described by a combination of Gaussian and Lorentzian functions, hence, the Pseudo-Voigt function enables one to mix the two functions in a η to ($1-\eta$) ratio. Further details on the relationships between these four functions and H, the half-width-at half-maximum value, are described by the Cagliotti function (*13*).

Once the the instrumental broadening and wavelength dispersion portions of the peak width are modeled, the remainder of the peak width can be described by the specimen function, which relates to the sample itself. More specifically, the peak broadening due to crystallite size and microstrain can be described mathematically. Scherrer's equation expresses peak broadening, β, as $\beta = \dfrac{\lambda}{\tau \cos\theta}$ where β is in units of radians, λ is the wavelength of the incident X-ray beam, θ is the diffraction angle, and τ is the crystallite size (*4*). It is important to note that this equation typically applies to samples with crystallite sizes on the order of 500 Å is accurate within an order of magnitude (*14*).

Diffraction and Nanostructures

Samples with particles below 100 nm provide challenges to the typical diffraction experiment. For example, alloys, defects and lack of long-range order in nanoparticle samples result in smearing of Bragg peaks and failed analyses.

Powder Diffraction Analysis

Samples of nanocrystalline powders result in complex peak broadening and require more more sophisticated analyses of diffraction peak widths. Cervellino et al. have recently developed full powder diffraction pattern calculations of nanoparticles based on log-normal size distribution of spherical particles(*15*). Balzar et al have also demonstrated with a carefully prepared nanocrystalline CeO_2 powder sample, that both size and strain contribute to peak broadening (*16*). It is suggested that Rietveld analysis is perhaps the best technique for samples where the experimenter does not have *a priori* knowledge of sample size and strain; however, the most significant drawback is the correlation of strain to other refineable parameters. Thus, both Lorentzian and Gaussian contributions to peak shape must be considered in nanoparticle samples.

Small-Angle X-Ray Scattering

Nanostructures are often disordered on a short length scale, yet ordered on a longer length scale of hundreds of Ångstroms rather than tens of Ångstroms. Application of Bragg's law to this distance implies that with Cu Kα radiation, the diffraction angle, θ, occurs at 0.45°. Thus, the particle shape, packing density, and heterogeneities of nanostructures can be explored using small angle X-ray scattering techniques, which were initially developed to study colloids, polymers, and micelles (14, 17). Recent applications of small angle scattering include structural defects, magnetic inhomogeneities, nanoscale porosity, and structure of thin films and membranes (18).

Scattering at small angles results in intense, continuous scattering with no clear, distinguishable Bragg peaks. This continuous scattering, first observed by Krishamurti (19) and Warren (20), results from the fact that the incident X-ray is in-phase and nearly parallel to the scattered ray. As the 2θ angle increases, the phase difference between two progressing waves also increases and scattering intensity decreases. This relationship between angle and scattering intensity applies to particle sizes that satisfy the relationship, $2\theta = \lambda/D$, where D is the particle size. If D is on the order of several wavelengths, λ, the scattering is spread over a wide range of θ, but intensity will be weak; if D is too large, the 2θ range will be physically inaccessible by the instrument. A typical 2θ range for experiments on nanoparticles is between 0.10° - 10°.

Data analysis of small angle-scattering X-ray and neutron data enables one to obtain information on particle shape, packing density, and heterogeneities. The reader is encouraged to refer to other references on this topic, as this is outside the scope of this paper.

References

1. Larson, A. C.; Von Dreele, R. B. *General Structure Analysis System (GSAS)*, 2000.
2. Pecharsky, V. K.; Zavalij, P. Y. *Fundamentals of Powder Diffraction and Structural Characterization of Materials*; Kluwer Academic Publishers: Boston, 2003.
3. Stout, G. H.; Jensen, L. H. *X-Ray Structure Determination: A Practical Guide*; 4th ed.; The Macmillan Company: New York, 1968.
4. Cullity, B. D.; Stock, S. R. *Elements of X-Ray Diffraction*; 3rd ed.; Prentice Hall: Upper Saddle River, 2001.

5. Bragg, W. L., *Proceedings of the Cambridge Philosophical Society* **1914**, *17*, 43-57.
6. Ewald, P. O., *Z. Kristallogr.* **1921**, *56*, 129.
7. The Powder Diffraction File is maintained by ICDD - International Centre for Diffraction Data http://www.icdd.com. The Inorganic Crystal Structure Database is produced by the Fachinformationszentrum Karlsruhe (FIZ) http://www.fiz-karlsruhe.de/and the National Institute of Standards and Technology (NIST): http://www.nist.gov/srd/nist84.htm
8. Shirley, R., Accuracy in Powder Diffraction. In *NBS Spec. Publ.*, Block, S. H., C.R., Ed. 1980; Vol. 567, pp 361-382.
9. Toby, B. H., *J. Appl. Crystallogr.* **2001**, *34*, 210-213.
10. Hunter, B. *Rietica - A visual Rietveld program*; 1998.
11. Rietveld, H. M., *Acta Cryst.* **1967**, *22*, 151.
12. Rietveld, H. M., *J. Appl. Crystallogr.* **1969**, *2*, 65.
13. Cagliotti, G.; Paoletti, A.; Ricci, F. P., *Nucl. Instrum. Methods* **1958**, *3*, 223.
14. Guinier, A.; Fournet, G., *Small-Angle Scattering of X-Rays*. John Wiley and Sons, Inc.: New York, 1955.
15. Cervellino, A.; Giannini, C.; Guagliardi, A.; Ladisa, M., *Physical Review B* **2005**, *72*, 9.
16. Balzar, D.; Audebrand, N.; Daymond, M. R.; Fitch, A.; Hewat, A.; Langford, J. I.; LeBail, A.; Louër, D.; Masson, O.; McCowan, C. N.; Popa, N. C.; Stephens, P. W.; Toby, B. H., *J. Appl. Crystallogr.* **2004**, *37*, 911-924.
17. Glatter, O.; Kratky, O., *Small Angle X-Ray Scattering*. Academic Press: New York, 1982.
18. An excellent review paper discusses recent advances in small angle X-ray and neutron scattering with respect to ceramic materials.
19. Krishnamurti, P., *Indian J. Physics* **1929-1930**, *3*, 507-522.
20. Warren, B. E., *J. Chem. Phys.* **1934**, *2*, 551-555.
21. Kraus, W.; Nolze, G. *PowderCell for Windows Verson* 1.1; Berlin, 1997.

Chapter 7

Development of Hands-on Nanotechnology Content Materials: Undergraduate Chemistry and Beyond

Sarah C. Larsen, Norbert J. Pienta, Russell G. Larsen

Department of Chemistry, University of Iowa, Iowa City, IA 52242

> Hands-on nanotechnology content materials based on semiconductor quantum dot nanocrystals were developed for use in a broad range of educational settings including undergraduate major and non-major chemistry courses, an upper level undergraduate chemistry laboratory course and outreach activities for K-12 and general public audiences. The focus is the colorful trend exhibited by cadmium selenide (CdSe) and cadmium selenide sulfide ($CdSe_xS_{1-x}$) quantum dots, which is correlated with the physical size of nanosized quantum dots. This visual trend illustrates one of the central themes in nanoscience that physical properties often depend on size at the nanoscale.

Introduction

Nanoscience is the study of the world on the nanometer scale, from approximately one to several hundred nanometers ([1]). One nanometer (nm) is one billionth of a meter. To put this into perspective, a human hair is approximately 50,000 nm in diameter, the individual components of computer chips are about 65 nm in size and a DNA molecule is 2.5 nm wide. *Nanotechnology* is the manipulation of molecules and atoms on the nanoscale and fabrication of nanoscale devices. The potential benefits of nanoscience and nanotechnology span diverse areas, including nanoelectronics, medicine, the environment, chemical and pharmaceutical industries, agriculture, biotechnology and computation.

© 2009 American Chemical Society

Given the importance of nanoscience on the scientific roadmap and the ever growing impact of nanotechnology on society, it is imperative to provide suitable education in nanoscience for all those that will need to navigate the societal changes that result. Since nanotechnology is expected to contribute to a diverse set of applications, education at all levels is required-graduate, undergraduate, K-12 students and the general public. The goal is to educate not only the future nanotechnology workforce, but to provide the general population with the scientific background necessary to make informed decisions related to nanotechnology. For these reasons, several different groups of students and lifelong learners are the focus of the materials being developed: (1) students who are likely to pursue a career in the physical sciences, engineering or medicine, (2) students who are non-science majors, (3) outreach audiences (including K-12 and adult).

Our goal was to develop hands-on activities and laboratory experiments that could be adapted for delivery to the different constituent groups (2). Semiconductor quantum dots were chosen as the intellectual focus of the new instructional materials. More specifically, colloidal quantum dots, formed from cadmium selenide (CdSe) and cadmium selenide sulfide ($CdSe_xS_{1-x}$) alloys were used to demonstrate the colorful trends that are correlated to the physical size of nanosized crystallites as shown in Figure 1. This visual trend vividly illustrates one of the central themes in nanoscience–*physical properties often depend on*

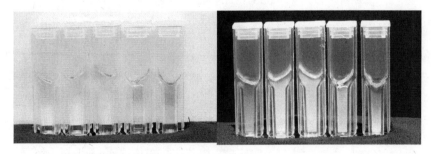

Figure 1. Nanometer-sized cadmium selenide sulfide ($CdSe_xS_{1-x}$) and cadmium selenide (CdSe). When the samples are irradiated with ultraviolet light ($\lambda = 365$ nm), the emission colors range from violet to orange (right). (See page 2 of color inserts.)

physical size on the nanoscale. For example, colloidal suspensions of CdSe with a narrow size distribution display colors ranging from a nearly colorless pale yellow for small crystallites (2 nm) to reddish (~7 nm) when viewed under white light. Colors due to photoluminescence range from violet to red-orange when viewed under long wavelength ultraviolet light (3). The $CdSe_xS_{1-x}$ system displays a similar trend but with an overall shift in the colors of light to shorter wavelengths (4, 5).

To develop the quantum dot laboratory and outreach materials, we concentrated on three key aspects of the visual trends observed for the quantum dots.

1. The visual features and empirical correlations that exist between observable optical properties and nanometer-sized crystallite dimension and composition of the quantum dots are ideal features for inquiry-based hands-on activities based on observable trends.
2. An empirical relationship between the visual colors and the physical size of the crystallites illustrates one of the fundamental ideas of nanoscience–that nanoscale materials display unique, size-dependent properties.
3. Intellectually, the key concepts that are necessary to understand these systems substantially overlap with pre-existing content within the chemistry courses that we are targeting. Concepts describing the nature of light and quantum nature of matter are some of the most important ideas in developing a modern view of chemistry.

In adapting materials to various audiences, the background, interests and abilities of the target groups must be considered and used to define and guide the adaptation process. Similarly, the environment into which the activites will be incorporated must also be considered. Our initial target was the large public university settings. We considered the chemistry course offerings at the University of Iowa, which has a total undergraduate enrollment of approximately 20,000. Our initial efforts focused on two of the chemistry courses at the most basic level of the curriculum. A one-semester general education course (3 credit + 1 credit optional lab) designed to examine the role of technology in society and a two semester general chemistry sequence (3 credit + 1 credit required lab each semester). These two courses serve large, non-overlapping populations of students. Further, these courses utilize up to 10 faculty per year and numerous graduate student teaching assistants. Therefore, the experience that these teaching professionals develop as they use the designed materials can be expected to have a multiplicative effect on the long term implementation and impact of nanoscience within the chemistry curriculum.

Two secondary targets were then identified–a First-Year seminar course and a physical chemistry laboratory course. These courses have much smaller and more selective enrollments but are important complements to broad coverage provided in the other two courses. Outside of the university setting, we have developed materials for use in a broad range of outreach activities for both school age and adult audiences.

Overall, five unique audiences are being addressed with one central set of nanotechnology content materials that have been adapted to each specific group:

1. A First-Year seminar course (enrollment limited to 15 or fewer). These courses are offered to undergraduates in their first year of matriculation on topics selected by the instructors;
2. A general education course fulfilling a laboratory science requirement, Technology and Society laboratory (4:006), taken by non-science majors, with an annual enrollment of 100-300;
3. An introductory chemistry course, Principles of Chemistry I (4:011), intended for science, engineering, and pre-professional majors with an annual enrollment of ca. 1500.

4. A required course for chemistry, biochemistry and chemical engineering majors, Physical Chemistry Laboratory (4:144), with an annual enrollment of ca. 25.
5. Outreach to K-12 and general public audiences

The general education and general chemistry courses together serve approximately 1800 students annually, representing about 35% of a typical entering class of undergraduates at The University of Iowa. Therefore, with this strategy we impact a sizable fraction of the student enrollment, cover a diverse range of interests, and include many of the targeted populations, at a critical juncture in their career development. In the context of these audiences, the next section outlines the adaptation strategies that enable the use of these core materials in all of these educational settings.

Upward Integration of Nanotechnology into the Undergraduate Chemistry Curriculum

Introductory Chemistry for Science Majors (Principles of Chemistry I)

Recently, the introductory chemistry courses, Principles of Chemistry I and II, at the University of Iowa, were redesigned to incorporate more active learning methods. Central to the curriculum redesign project was the restructuring of the laboratory component of the courses. Specifically, the laboratory, which had previously been offered as a separate course, was integrated with the lecture aspect of the course, while simultaneously creating a new course-component, the case study. In alternating weeks, students attend case study (80 minutes, week 1) and laboratory (170 minutes, week 2). In the case study, students complete activities and observe demonstrations designed to provide the background context of the topic. A case study and laboratory experiment based on semiconductor quantum dots were developed for the Principles of Chemistry I course (6).

Case Study

In the case study of quantum dots, a brief introduction is used to highlight nanotechnology, and to establish the basis for the investigation of the optical properties of quantum dots. As part of the introduction, each student is given a set of quantum dot samples, solidified in a polymer resin. Then, through observation and reasoning, the students develop an understanding of the optical properties of the nanosized semiconducting materials. Initially, students are reminded of useful conceptual tools introduced during lecture and how these tools can be used to interpret the observations made during the case study session. Prior to case study, students have been introduced to the ideas of visible light (including properties such as color, wavelength, and energy), and the mathematical relationships between wavelength and energy. During the case study session, the students explore the interaction of light with matter, the visual and spectroscopic measurements of light, and the quantum mechanics of energy levels and the absorption and emission of light. The quantum dot samples are used to illustrate these concepts.

Next, the students actively observe the luminescent properties of the quantum dot samples. The lights are turned down and the students are instructed to shine the light across the quantum dot samples. For the provided samples, the fluorescent colors range from violet to yellow-orange depending on the sample viewed by the students. The instructor has a complete set, including violet, blue, blue-green, green, green-yellow, yellow, and yellow-orange samples. The students are told that this is the observation that they are to understand in the next part of the activity.

Viewed under white light, the polymerized quantum dot samples exhibit a trend in color ranging from very pale yellow to orange. The students are asked to consider the most extreme colors and are presented with pictures of samples of these two cases. The students are then presented with two absorption spectra and asked to match the spectrum to the corresponding sample. The students construct energy level diagrams that are consistent with the absorption spectrum for each of the two samples. Generally, the energy diagram of the first sample is constructed as a class with prompting from the instructor and the diagram corresponding to the second sample is constructed by the students as part of their written work. Overall, one of the most important points is that the absorption onset is directly related to the band gap energy on the energy level diagram.

Students are presented with pictures of the same two samples fluorescing due to band-edge emission. One sample appears blue, whereas the second appears yellow-orange. Students are asked to match the sample to the corresponding absorption spectrum, using the described process of fluorescence in which the electron relaxes to near the bottom of the conduction band before undergoing a transition. Upon reflection, discussion, (and sometimes scaffolded questioning), students eventually use their constructed energy diagrams to realize that the emission should take place at the same wavelength as the absorption onset, since in the simple model, both correspond to the energy of the band gap.

Finally, the students are reminded that they have seen that band gap energies apparently change, but without any evidence of the reason that causes these observed changes. The instructor explains that there are at least two effective strategies to adjust the band gap of semiconducting materials. The first one is by changing composition (4), a strategy often used in adjusting the colors of LEDs (7). For example, bulk CdSe is known to have a band gap energy of 2.8×10^{-19} J, while CdS has a greater band gap energy of 3.9×10^{-19} J. It is also known that alloys formed as combinations typically show band gap energies with intermediate values (8). In lecture, students have looked at periodicity, and they easily accept the idea that electronic energies depend on chemical composition.

The second strategy is one that is newer to science and involves the nano-scale. The students are told that a key empirical finding is that the color of the different samples is a function of the size of the crystallites of semiconducting material that form the samples. The students are shown Transmission Electron

Microscope (TEM) images from representative CdSe samples and illustrations to suggest the relationship between size and a corresponding energy level diagram.

Laboratory Experiment

In the laboratory which takes place the week after the case study activity, students synthesize quantum dots and investigate quantum dot optical properties. Many logistical challenges (including large student numbers, a limited equipment budget, chemical toxicity, high temperature reaction conditions, and time limitations) were overcome in order to implement this experiment in a course this size.

In 2002, Rosenthal and co-workers reported using semiconductor nanocrystals as visual aides for introducing the particle-in-a-box concept to juniors and seniors in physical chemistry (*9*). However, at that time, the best chemical procedure for producing large quantities of high quality quantum dots was based on the work of Murray (*10, 11*). In this method, dimethyl cadmium, an extremely toxic and pyrophoric substance, was reacted at high temperatures under an inert atmosphere. The synthesis was expensive, difficult to control, and could potentially become explosive, certainly not a method suitable for an introductory laboratory with novice laboratory students.

A breakthrough that opened the door for the use of quantum dot synthesis in the teaching laboratories was the report of the air stable "greener" synthesis of Peng and co-workers (*3, 12, 13*.) This method had the advantages that it used lower temperatures, used a safer solvent system, could be carried out in air, used fewer hazardous starting reagents, and routinely produced relatively monodisperse samples with high luminescent quantum yields. These methods have also been well characterized; therefore, the extinction coefficient and absorption peaks correlate well with TEM measurements of crystallite size (*3*).

One adaptation of this procedure for use in an instructional laboratory has recently been reported (*14*). This new procedure represents a successful method that conforms to the time constraints imposed by an experiment in an undergraduate setting, uses equipment typically found in a synthetic laboratory, and manages the safety issues so that use in an undergraduate laboratory class can be considered. However, implementation of these synthetic methods in a high enrollment, introductory setting remained a challenge due to the approximately 1500 students enrolled annually. Many practical issues enumerated below had to be recognized and resolved in order to institute the method into the introductory laboratories

To provide reliable, simultaneous heating of numerous samples, an aluminum culture plate on a magnetically stirred hotplate was used to heat the synthesis solutions. The temperature of the aluminum block can be monitored using a thermocouple in contact with the aluminum block. This provides safe, stirred heating of up to twelve samples simultaneously. To minimize exposure to chemicals, dilute stock solutions are provided to the student. The preparation of these dilute solutions use potentially hazardous chemicals including metallic selenium, trioctylphosphine or tributylphosphine, and cadmium oxide (*12-14*) that are easily managed by the professional preparatory room staff.

Furthermore, the concentration of the cadmium solution was reduced by a factor of four compared to the previous concentrations reported. This dilution offers the dual advantage of reducing the concentration of cadmium while also lowering the optical density of the solutions, thereby increasing the penetration depth of excitation source used in the measurement of the fluorescent emission spectra. As prepared, the dilute solutions have a working lifetime of greater than a week.

Students perform the syntheses in disposable test tubes (18x150 mm), outfitted with a small magnetic stir bar. Each student transfers three milliliters of cadmium stock [containing 1. mg (7.6 μmol) CdO in 3 mL of 6% oleic acid in 1-octadecene] into the disposable test tube using a disposable pipet. The solution is positioned in the aluminum block and allowed to equilibrate for 5 minutes to a temperature of 220° C. To prepare CdSe quantum dots, the students rapidly transfer 0.5 mL of room temperature selenium stock solution [containing 25 μmol of Se, 33 μmol of the phosphine, and 0.5 mL 1-octadecene](12-15) into the test tube. Students are grouped in teams of four, and each team member is responsible for a different reaction time between 0.25 and 3 minutes. Once the allotted time has elapsed, the test tube is removed and positioned into a room temperature aluminum block to cool. After the sample has cooled, the students stopper their test tubes to prevent accidental spilling, record the trends in the colors observed under ambient light and under irradiation using a 400 nm violet penlight (or long wavelength ultraviolet light). Other groups of four, prepare $CdSe_{0.5}S_{0.5}$ using an analogous procedure in which a selenium/sulfur stock solution [containing 12 μmol of Se, 12 μmol of S, 16 μmol of the phosphine, and 0.5 mL 1-octadecene] initiates the reaction.

Student prepared CdSe samples display colors under ambient light ranging from pale yellow to orange (and from blue-green to yellow-orange under 400 nm excitation)(14) and $CdSe_{0.5}S_{0.5}$ samples that range from nearly colorless to yellow under ambient light (and violet to green under 400 nm excitation). In general, the $CdSe_{0.5}S_{0.5}$ samples display brighter fluorescence and greatly facilitate preparation of samples with fluorescent colors in the violet to blue-green range.

Since students share the fume hood during the synthesis of their quantum dot samples, we have elected to provide samples to use for spectroscopic analysis. This has several advantages: (1) it allows the students to efficiently use their time by decoupling the time needed for the synthesis from characterization, enabling them to be conducted in parallel; (2) it eliminates the need to transfer the sample from the large reaction test tubes to the small test tubes used for the optical measurements, and reduces the potential chemical exposure that is inherent in each chemical manipulation; (3) it enables samples that have been solidified in a polymer resin to be used to eliminate the possibility of spilling; and (4) it assures that samples of known quality are used for the spectroscopic analysis. This last advantage should not be undervalued, since the expected conceptual ties are critically tied to the "correct" empirical trends, and although the experimental methods are robust, they are not foolproof.

The polymerized samples for case study sessions and for use during spectroscopic measurements in the laboratory are prepared by the laboratory staff, using reaction mixtures containing colloidal quantum dots. First, samples of quantum dot are prepared as described above except that the cadmium oxide stock solution is four times more concentrated than the stock used by the student. In other words, similar conditions are used to those in ref. (*14*). Reaction times are varied to produce samples covering as broad of a range of colors as possible for both the CdSe and $CdSe_{0.5}S_{0.5}$ systems. After cooling, the samples are each diluted 1:3 (volume of colloid sample to volume of methacrylate mixture) with a 5% ethylene dimethacrylate in lauryl methacrylate with 0.5% benzoin methyl ether added as a photoinitiator. The methacrylates were used as provided by the manufacturers and include the stabilizer. Aliquots of the mixture are distributed in either small vials for case study sessions or semi-microcuvettes (or test tubes) for spectroscopic measurement. The samples are irradiated with long wave UV (~366 nm) from a handheld lamp for 0.75-4 hours, resulting in a solid sample sufficiently clear for spectroscopic study. Samples prepared in this way display some variation in lifetime, and typically lasted from 1 month to greater than a year with sufficient fluorescence for visual and spectroscopic detection. Additional work is in progress to determine the factors that govern sample lifetime in these samples. Although commercial samples of polymer-embedded colloidal quantum dots are available, the current price of these high quality samples prohibits their use in most instructional settings that require hundreds of samples. Therefore, the cost and simplicity of this sample preparation method is an important advance in making these materials accessible to students in hands-on classroom settings.

The size of the quantum dots in the prepared samples is estimated based on the empirical correlations of Peng,(*3*) and each sample is labeled with its corresponding estimated size. For the $CdSe_xS_{1-x}$ alloy, an estimate can be made based on the known value of CdSe, CdS and the alloy (*5*). We assume that for a given crystallite size, the average (weighted according to composition) of the two equations would be obtained, and this is viewed as reasonable since band bowing is known to be small in comparable bulk alloy systems (*8*).

Students measure a set of samples to establish the relationship between the labeled size, the absorption onset (*16*) and the emission maximum. From their case study preparation, students expect that both measurements should be an approximate measure of the effective band gap of these systems. For such systems, the particle-in-the-box model suggests that a plot of effective band gap energy versus the reciprocal of the sample dimension squared should yield a straight line with an intercept approximately equal to the band gap energy of the bulk. The empirical relationship reported by Peng (*3*) relates the wavelength of the first absorption maxima to the TEM-measured size of CdSe nanocrystals. This together with the observation by Boatman (*14*) that the emission maximum is displaced by approximately 25 nm to longer wavelength compared to the absorption maxima gives a combined empirical relationship. Within this range, the data are largely compatible with this strong confinement model and elucidated by the particle-in-the-box model. A more refined model, appropriate for more advanced students, would explicitly consider both the effective masses of the electrons and holes and the Coulombic interaction energy (*16-18*).

In our laboratories, students use CCD-spectrometers to measure the absorbance of the samples and graphically find the absorption onset. The students also measure the fluorescence of these samples using the same spectrometer (with the light source off) by exciting the sample down the sample axis (orthogonal to detection) with a violet LED penlight (~400 nm). Due to the high quantum yield in these systems, the emission maxima are easily identified by this method. For institutions that do not have CCD-spectrometers widely available for student use, a Spectronic-20 or similar spectrometer could be used to measure the absorbance onset. Although, the spectral range of the fluorescence emission can be visually measured either by color-wavelength correlations or by using a hand held Project Star Spectrometer, the loss of quantitative intensity information limits the conclusions that can be drawn from such visual observations. However, as long as the absorbance onset can be measured, the above linear relationship will be returned by careful measurement, since the quantum dot dimension has been assigned using the established empirical relationships.

Chemistry for Non-majors (Chemistry and Technology)

Technology and Society (4:005) is a one semester three credit hour chemistry course for non-science majors that partially fulfills the general education requirement in science at the University of Iowa. There is also a one credit hour laboratory (4:006) in which students have the option of enrolling concurrently. The *Chemistry in Context* textbook (*19*) and laboratory manual (*20*) are currently used in 4:005 and 4:006, respectively. Our general strategy was to adapt the quantum dot instructional materials that were used in the Principles of Chemistry I course, recognizing that the course objectives and the backgrounds of the students in the course are quite different.

Since the laboratory course is independent of the lecture course and since the course does not include a laboratory lecture or case study session, written material in the form of a pre-laboratory handout is used to introduce nanotechnology and highlight. Students were reminded of the related ideas about the properties of light that they had already covered in the lecture portion of the course.

Students conducted four main activities during the two hours allotted to this laboratory:
1. A sample of colloidal quantum dots was synthesized by each student.
2. Visual observations of the apparent colors (under ambient light and violet light) of a pre-prepared, polymerized set of quantum dot samples were recorded.
3. Measurements of the absorption and emission spectra of the pre-prepared samples were made and the wavelength values of the absorption and emission maxima were recorded.
4. Plots were constructed relating the size of the quantum dots to the measured emission maxima and absorption onset wavelengths.

It should be noted that the laboratory activities are similar to the Principles of Chemistry laboratory experiment with the most significant difference being that the quantum mechanical aspects of the quantum dots were not developed in the Technology and Society course.

First-Year Seminar Course

At the University of Iowa, a First-Year Seminar program is offered to first and second semester undergraduate students who may enroll for one credit hour in a seminar course that provides them with an opportunity to work closely with a faculty member. The topic is chosen by the faculty member and is often related to the faculty member's research. First-Year Seminars are graded on participation and written papers, and class sizes are limited to 16 students.

Using this venue, a First-Year Seminar course entitled *Explorations in Nanoscience and Nanotechnology* was offered at the University of Iowa in Spring 2003 and Fall 2005. In this seminar course, the fundamental concepts of nanoscience were introduced, the properties of nanoscale materials were examined, and applications in nanotechnology were presented in a series of laboratory experiments and hands-on demonstrations. The first unit of the course included several laboratory explorations involving nanomaterials, such as memory metal, LED's (*21*), ferrofluids (*22*) and quantum dots. The quantum dot experiment was adapted for this course from the course materials used for 4:011 Principles of Chemistry and 4:006 Technology and Society Laboratory previously described in this chapter.

Because of time constraints for the course period, the optical properties of quantum dots were investigated using pre-prepared CdSe samples with sizes ranging from 2 to 4 nm. The absorption and luminescence properties of the quantum dots and the dependence of these properties on the size of the quantum dots were examined by the students in a manner similar to the experiment developed for the Technology and Society laboratory. In the second unit, students were introduced to the tools of nanoscience through a tour of the Central Microscopy Facility at the University of Iowa and an experiment utilizing scanning tunneling microscopy. The third unit focused on applications of nanoscience and included an investigation of the properties of LCD materials, the synthesis of gold nanoparticles (*23*), and the construction of a nanocrystalline TiO_2 solar cell. In Fall 2005, an additional unit related to the societal implications of nanoscience and nanotechnology was added to the course. Courses such as this are complementary to efforts to incorporate nanotechnology in the mainstream courses, because they offer a broader coverage of topics in nanotechnology for those students that are interested but are not necessarily science majors.

Upper Level Undergraduate Laboratory Courses (Physical Measurements)

Although physical chemistry was not among our primary targets and has been treated by others in the literature (*14, 16, 24, 25*), the influence of the intro-level project made an impact on the physical chemistry laboratory course. In the fall of 2005, the instructor of the physical chemistry course was aware of the quantum dot project and contacted those involved in the freshman project to see if samples could be provided for the physical chemistry course. The instructor was primarily interested in having students study quantum dot samples using a fluorimeter. Since the focus was intended to be on the spectroscopy and theoretical quantum mechanics that are useful in interpreting the results, some colloidal solutions of quantum dots were provided to the instructor and were used by the students in the course. In future semesters, such samples may literally be synthesized in the freshman laboratories for analysis in the advanced laboratories, providing an interesting upward integration into the curriculum.

Development of "Nano-to-Go" Quantum Dots

The quantum dot activities originally developed for use in undergraduate chemistry laboratories were found to be very adaptable and could be modified for a variety of educational settings involving hands-on activities. In particular, the polymerized quantum dot samples are inexpensive to prepare, safe to use, self-contained, spillproof, and easily transportable. As we have expanded our outreach efforts to include K-12 audiences and general public audiences, we have developed "Nano-to-go" activities which can easily be incorporated into presentations suitable for many different outreach groups. A card bearing the NANO@IOWA logo at the top and a black background at the bottom holds the polymerized quantum dot samples. A violet LED pen light is attached with a string to the quantum dot card as pictured in Figure 2. A presentation describing the fundamental aspects of nanoscience accompanies the "Nano-to-go" quantum dots. These quantum dot samples and penlights are passed out to the audience as the presenter discusses size-dependent properties of nanomaterials to reinforce the concepts. The "Nano-to-go" quantum dots have been used in a wide-range of outreach venues including middle school science students and university alumni groups.

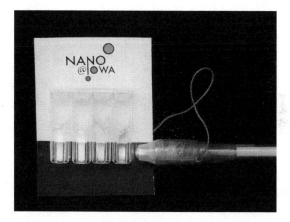

Figure 2. Cadmium selenide (CdSe) quantum dot samples embedded in a polymer are used for outreach activities. The luminescent properties can be observed using the attached violet LED pen lights. The NANO@IOWA logo on the cards reinforces the idea that the colors of the quantum dots depend on their size.
(See page 3 of color inserts.)

Conclusions

To summarize, nanoscience curricular materials based on semiconductor quantum dots were adapted for both formal and informal educational purposes.

The technical level depends on the audience which ranged from middle school students to upper level undergraduate chemistry majors in a physical chemistry laboratory course. For outreach and non-science major chemistry courses, pre-prepared quantum dot samples immobilized in a polymer matrix were used to illustrate the size dependence of the optical properties of CdSe and $CdSe_xS_{1-x}$ alloys. For introductory chemistry courses, students synthesized CdSe and $CdSe_xS_{1-x}$ alloy samples in addition to examining the optical properties of the quantum dots. The topic of semiconductor quantum dots was chosen to engage the students in order to generate interest and maintain motivation using the observable and colorful trend with size.

Acknowledgements

Undergraduate students, Amanda Drevyanko, Rachelle Justice and Jamie O'Meara and high school student, Haley Lamm are acknowledged for their participation in the project as research assistants. Financial support from the National Science Foundation (NUE-CHE0407162) is acknowledged. Drs. Dale Swenson, Mark Young, Mohammed El-Maazawi and John Kirk are acknowledged for assistance with implementation of educational materials in other courses.

References

1. Ratner, M.; Ratner, D. *Nanotechnology: A Gentle Introduction to the Next Big Idea*; Prentice-Hall: New Jersey, 2003.
2. Larsen, R. G.; Pienta, N. J.; Larsen, S. C. In *Nanoscale Science and Engineering Education: Issues, Trends and Future Directions*; Seal, D. A. E. S. a. D. S., Ed.; American Scientific Publishers, Stevenson Ranch, CA 2008.
3. Yu, W. W.; Qu, L.; Guo, W.; Peng, X. *Chem. Mater.* **2004**, *16*, 560.
4. Bailey, R. E.; Nie, S. *J. Am. Chem. Soc.* **2003**, *125*, 7100.
5. Swafford, L. A.; Weigand, L. A.; Bowers, M. J.; McBride, J. R.; Rapaport, J. L.; Watt, T. L.; Dixit, S. K.; Feldman, L.; Rosenthal, S. J. *J. Am Chem. Soc.* **2006**, *128*, 12299.
6. Larsen, R. G.; Tardy, D.; Haverhals, L.; Cannon, L.; Pienta, N. J.; Larsen, S. *(Course-pak, Iowa City, IA, 2004)* **2004**.
7. Lisensky, G. C.; Penn, R. *J. Chem. Educ.* **1992**, *69*, 151.
8. Wei, S.-H.; Zhang, S. B.; Zunger, A. *J. Appl. Phys.* **2000**, *87*, 1304.
9. Kippeny, T.; Swafford, L. A.; Rosenthal, S. J. *J. Chem. Educ.* **2002**, *79*, 1094.
10. Murray, C. B.; Norris, D. J.; Bawendi, M. G. *J. Am. Chem. Soc.* **1993**, *115*, 8706.
11. Crouch, D.; Norager, S.; O'Brien, P.; Park, J.-H.; Pickett, N. *Phil. Trans. R. Soc. Lond. A* **2003**, 297.
12. Qu, L.; Peng, X. *J. Am. Chem. Soc.* **2002**, *124*, 2049.
13. Yu, W. W.; Peng, X. *Angew. Chem. Int. Ed.* **2002**, *41*, 2368.

14. Boatman, E. M.; Lisensky, G. C.; Nordell, K. J. *J. Chem. Educ.* **2005**, *82*, 1697.
15. Li, J. J.; Wang, Y. A.; Guo, W.; Keay, J. C.; Mishima, T. D.; Johnson, M. B.; Peng, X. *J. Am Chem. Soc.* **2003**, *125*, 12567.
16. Nedeljkovic, J. M.; Patel, R. C.; Kaufman, P.; Joyce-Pruden, C.; O'Leary, N. *J. Chem. Educ.* **1993**, *70*, 342.
17. Brus, L. E. *J. Chem. Phys.* **1983**, *79*, 5566.
18. Brus, L. E. *J. Chem. Phys.* **1984**, *80*, 4403.
19. *Chemistry in Context- Applying Chemistry to Society*; 4th ed.; Stanitski, C. L.; Eubanks, L. P.; Middlecamp, C. H.; Pienta, N. J., Eds.; McGraw-Hill: New York, 2003.
20. *Laboratory Manual, Chemistry in Context- Applying Chemistry to Society*; 4th ed.; Stratton, W. J.; Steehler, G. A.; Pienta, N. J.; Middlecamp, C. H., Eds.; McGraw-Hill: New York, 2003.
21. Condren, M. S.; Lisensky, G. C.; Ellis, A. B.; Nordell, K. J.; Kuech, T. F.; Stockman, S. A. *J. Chem. Educ.* **2001**, *78*, 1033.
22. Berger, P.; Adelman, N. B.; Beckman, K. J.; Campbell, D. J.; Ellis, A. B.; Lisensky, G. C. *J. Chem. Educ.* **1999**, *76*, 943.
23. McFarland, A. D.; Haynes, C. L.; Mirkin, C. A.; Duyne, R. P. V.; Godwin, H. A. *J. Chem. Educ.* **2004**, *81*, 544A.
24. Kippeny, T.; Swafford, L. A.; Rosenthal, S. J. *J. Chem. Educ.* **2002**, *79*, 1094.
25. Winkler, L. D.; Arceo, J. F.; Hughes, W. C.; DeGraff, B. A.; Augustine, B. H. *J. Chem. Educ.* **2005**, *82*, 1700.

Chapter 8

Introducing Nanotechnology into Environmental Engineering Curriculum

X. Zhang[1], C. Bruell[1], Y. Yin[1], P. Jayarudu[1], and A. Watterson[2]

[1]Department of Civil & Environmental Engineering
[2]Department of Chemistry
University of Massachusetts Lowell
Lowell, MA 01854

Three laboratory modules addressing the environmental impacts of nanotechnology were developed and implemented into one junior-level undergraduate environmental engineering course. Detailed laboratory procedures for each module are presented together with student results. All modules were fully evaluated with two survey instruments, feedback surveys and multiple choice pre and post-tests. Overall, students had a positive exposure to nanotechnology and such experience enhanced their learning and increased their interest in science and/or engineering.

Introduction

As a result of the National Nanotechnology Initiative (*1*), substantial advances have been made in using nanotechnology to generate nanomaterials with novel properties (*2*). These materials and processes have or will produce products ranging from coatings for stain resistant fabrics to smaller and faster computer chips. Recent studies have also suggested that nanotechnology can be employed in pollution prevention, treatment, and remediation (*2, 3*).

The use of commercially available zero-vale12nt-metal powders for the degradation of halogenated aliphatics is well documented (*4*). Nanoscale Fe^0 has a much smaller grain size than commercially available powdered iron, making it much more reactive. Nanoiron and nanoscale bimetallic particles have been shown to be extremely effective for the reductive dehalogenation of common soil and ground water contaminants such as: chlorinated methanes (*5*), chlorinated ethanes (*6*) and chlorinated ethenes (*7, 8*) and essentially eliminate all the undesirable byproducts (*9*).

© 2009 American Chemical Society

Lead is among the most toxic elements and has widespread presence in the environment (*10, 11*). Common treatment technologies for lead removal include chemical precipitation and adsorption. However, precipitation becomes less effective and more expensive at high metal concentrations (*12*) and successful adsorption depends on finding low-cost, high-capacity sorbents (*12-23*) or microorganisms that accumulate toxic metals (*24-26*). Innovative nanospheres have shown promise for lead complexation.

The academic community has reacted to offer either graduate or undergraduate courses in nanoscicence or nanoengineering (*27, 28*). However, there is very little effort to introduce nanotechnology into the undergraduate environmental engineering curriculum. The objective of this project was to introduce nanotechnology experiences into the undergraduate environmental curriculum so that students will be exposed to cutting-edge advances in nanotechnology and their impact on the environment.

Overview of Modules

Three research-based environmental nanotechnology modules have been designed and implemented in an undergraduate level environmental engineering course. For these modules, we selected two nanomaterials (nanoscale bimetallic iron particles and engineered nanospheres) that may provide solutions to challenging environmental pollution problems (*3*). Table I shows a summary of each module and its learning objective(s).

Module Creation and Description

<u>Module 1:</u> Synthesis of nanoscale bimetallic iron particles. The module has been created according to existing literature. Nanoscale iron particles are synthesized by mixing $NaBH_4$ (0.25 M) and $FeCl_3 \cdot 6H_2O$ (0.045 M) solutions (1:1 volume ratio) (*8*). The reaction is as follows:

$$4 Fe^{3+} + 3 BH_4^- + 9 H_2O \rightarrow 4 Fe^0 (s) + 3 H_2BO_3^- + 12 H^+ + 6H_2(g)$$

The freshly prepared nanoiron particles are then coated with palladium acetate to form Pd/Fe bimetallic particles (*8*).

Table I. Overview of Modules

Module	*Objective*
Module 1: Synthesis of Palladized Nanoscale Iron Particles	To understand chemical synthesis for environmental remediation
Module 2: TCE Degradation with Palladized Nanoscale Iron Particles	To apply nanotechnology for groundwater remediation
Module 3: Use of Engineered Nanospheres for Lead Complexation Lab	To determine the effectiveness of novel nanospheres for lead complexation

Module 2: Using nanoscale bimetallic iron particles for groundwater remediation. This module has been created according to existing literature. Trichloroethylene (TCE), one of the most ubiquitous soil and groundwater contaminants, is used as a sample contaminant. The reductive dehalogenation of TCE via zero-valent nano iron particles can be described by the following equation (29):

$$3Fe^{\circ} + C_2HCl_3 + 3H_2O \rightarrow C_2H_4 + 3\ Fe^{2+} + 3Cl^- + 3OH^-$$

However, the reaction of using palladized iron particles is not clear. TCE degradation by Pd/Fe nano-particles is assumed to follow pseudo-first-order kinetics. The pseudo-first-order decay coefficient and half life ($t_{1/2}$) can be determined after plotting experimental data.

Module 3: Using engineered nanospheres for lead removal. This module has been created in conjunction with ongoing research. This module evaluates the potential of using nanospheres functionalized with carboxyl groups for lead complexation from aqueous solutions. Bench-scale experiments are conducted to observe Pb^{2+} complexation after mixing carboxyl functionalized nanospheres with various concentrations of Pb^{2+} solution (6-30 mg/L) for 30 minutes. Residual free lead ion concentration is quantified by a lead ion selective probe. The average molecular weight of the carboxyl functionalized nanospheres is 12,000 g.

Course Description

Junior-level "Environmental Engineering (3 credit)" and "Environmental Engineering Laboratory (1 credit)" are core companion courses for Civil & Environmental Engineering students. The course and laboratory focus on the physical, chemical, and biological principles of water and wastewater treatment, the design of wastewater treatment facilities (WWTFs), and hazardous waste site remediation. The current experiments concentrate on measures for performance of WWTFs, such as biochemical oxygen demand, chemical oxygen demand, suspended solids, etc. The addition of Modules 1, 2 and 3 to the laboratory greatly expanded the existing curriculum and introduced students to the use of novel technologies for the solution of environmental problems.

Module Implementation

For each module implementation, a brief power point presentation was delivered by the instructor on the overall background of nanotechnology and the specifics of the module. Students were given a lab handout containing detailed background information and the laboratory procedure. Upon completion of each module experiment, each student wrote a laboratory report analyzing the data generated by the whole class, answered several questions related to each lab and provided a discussion of the experimental results.

- In Module 1, students were given the appropriate chemical solutions and they followed detailed procedures (see Appendix A) to synthesize Pd/Fe nanoiron particles.
- In Module 2, students monitored the removal of TCE using the Pd/Fe made in Module 1 (see Appendix B for the experimental procedure). TCE was measured by a gas chromatograph. The students obtained the following plot shown in Figure 1 and determined the kinetic constant for TCE degradation to be 0.0393 min^{-1} with a half life of 17.6 minutes.
- In Module 3, students were divided into four groups and each group was given the same amount of carboxyl functionalized nanospheres to complex varying initial concentrations of free lead ion (each group used a different initial Pb^{2+} concentration) (see Appendix C for the experimental procedure). Lead complexation efficiency was determined by the students using an ion selective probe. The class plotted the following graph (shown in Figure 2) using the molar ratio between the carboxyl functionalized nanospheres to initial free lead concentration vs. lead complexation efficiency. The students were asked to evaluate the impact of the amount of nanospheres used on lead complexation based on the results obtained from the class.

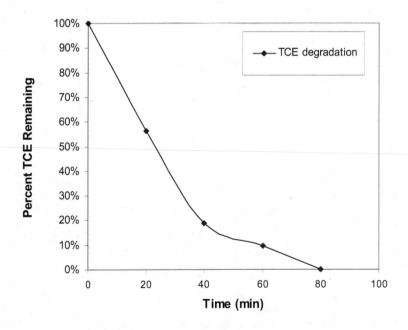

Figure 1. TCE dechlorination by nanoscale Pd/Fe.

Assessment

Modules were evaluated by the University of Massachusetts Donahue Institute (UMDI). Two survey instruments were used, namely feedback surveys (Appendix D) and multiple choice pre-post-tests (see Appendixes E, F, and G for specific pre- post- test questions for each module). They were distributed to students during class time. Of the twenty-six students who completed feedback surveys, 69% were male and 31% were female. Eighty-one percent of respondents were Caucasian, 12% were Hispanic/Latino and 8% were Asian. All students were taking the course as a requirement and most students (92%) were enrolled as juniors. Two senior students were also enrolled in the course. Forty-two percent of the students reported that they plan to attend graduate school. Approximately 12% do not plan to attend graduate school and 46% were not sure.

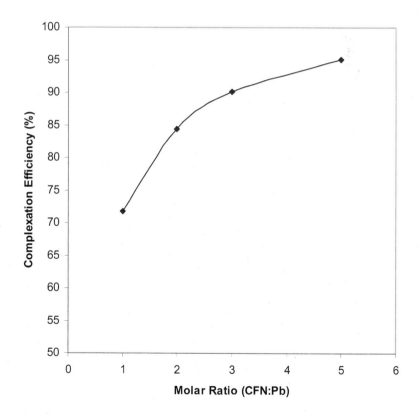

Figure 2. Lead complexation with carboxyl functionalized nanospheres.

Quality of the Modules

Quality of the modules was assessed by the feedback survey. Student responses to questions about the modules were positive overall.
- "I had a better understanding of nanotechnology after completing the module" (90%)
- "I would like to experience additional nanotechnology modules in other courses" (42%) Forty-six percent of students did not agree or disagreed with this statement.
- "My interest in science (and/or engineering) has increased as a result of the nanotechnology modules" (42%)
- "I am interested in conducting further research related to nanotechnology" (31%)

The final questions on the feedback surveys asked students to indicate what they liked best about the module, what they found most challenging, and any additional comments or suggestions that would improve the quality of the module and/or the students' learning experience. Twenty-four students provided information regarding what they liked about the module.
- "enjoyed learning about nanotechnology" (38%) Several of these students indicated that this was their first experience with nanotechnology and the new technology in general.
- "enjoyed the hands-on nature of performing the experiment" (29%)
- "liked the real life application of the experiment and/or that is was useful for environmental purposes" (21%)
- "liked it because it was "interesting," "neat," or "fun"" (21%)
- "liked the module because it was a relatively easy procedure" (8%)
- "enjoyed the application of a new and innovative technology" (8%)
- others include: "I liked making the iron for the TCE/Pd," "Learning how it is a useful study for current research," and "The process was fine tuned before the start of the lab, making it go smoother."

Twenty-one students responded to the question asking what they found most challenging about the module.
- "understanding the concepts and details related to the experiment was most challenging" (48%)
- "getting enough accurate data and results" (25%)
- "the required precision of the steps needed to conduct the experiment" (25%)

Comments and suggestions related to module improvement are listed below. They tended to focus on additional information and background related to nanotechnology and more applied analysis.
- "it would be interesting to do the whole lab (our TA did some of it because of time constraints)"
- "the procedure should be refined so that more efficient results are obtained"

- "more class time to better understand nanotechnology should be considered"
- ""Nano" is still a very intimidating/confusing term for me - defining and explaining it in a more succinct and clear way would have helped me"
- "previous practice of the techniques used, however this is an introduction to nanotechnology and we are not expected to have prior knowledge"
- "previous knowledge of nanotechnology before the lab and/or experience with nanotechnology"
- "more exposure to nanotechnology"

Student Learning

Student learning was assessed through the use of multiple-choice pre- and post-tests. For Modules 1 and 3, student scores on post-test questions were significantly higher after implementation of the modules, providing evidence that student learning increased as a result of the implementation of the modules.

Figures 3, 4, and 5 provide an analysis of the pre- and post-test results for Modules 1, 2 and 3, respectively. Each set of module data were also analyzed by comparing the percentage of correct responses to individual items on the pre-test versus post-testing. Of the eight questions used to assess student learning for Module 1, changes from pre- to post-testing were statistically significant for six of the questions. Students were able to answer the majority of the questions more accurately after implementation of the module, indicating an increase in student learning.

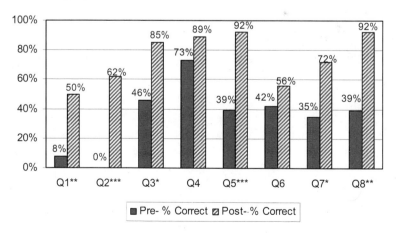

Figure 3. Pre- and post-test results by question for module 1. (Results are statistically significant * $p<0.05$, ** $p<0.01$, *** $p<0.001$)

Findings for module 2 indicated a decrease in the percentage of questions students answered correctly from pre- to post-testing for questions one and two, though it should be noted that these findings were not statistically significant. Questions three and four produced a slight increase in the percentage of correct responses from pre- to post-testing, but again these findings were not statistically significant. However, the increase in correct responses from pre- to post-test on questions five and six were statistically significant ($p<0.01$), suggesting that students had learned the material related to questions five and six based on module implementation. The decrease related to questions one and two may have been due to confusion related to the content material and/or wording of individual questions. Further examination of both module content and pre- post-test question wording may help to increase the likelihood of increasing student knowledge related to this module.

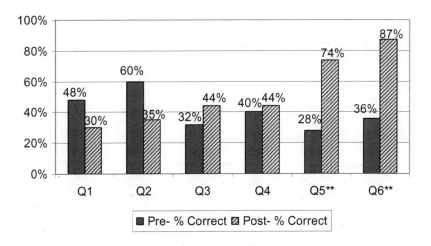

*Figure 4. Pre- and post-test results by question for module 2. (Results are statistically significant ** $p<0.01$)*

For module 3, the mean increase in the percentage of correct responses from pre- to post-testing was statistically significant ($p<0.001$). Students were only able to answer half of the questions correctly before the module was implemented and were able to answer almost 90% of the questions correctly after module implementation. An examination of the results by individual item also shows statistically significant gains in student learning related to each of the five questions.

The feedback surveys and pre-post tests scores from all three modules suggested that the students had positive exposure to nanotechnology and such experience enhanced their learning experience and their interest in science and/or engineering.

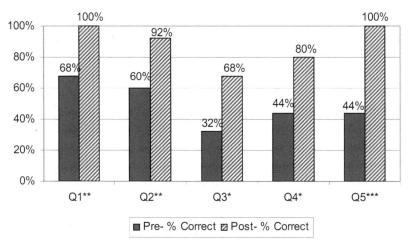

*Figure 5. Pre- and post-test results by question for module 3. (Results are statistically significant * $p<0.05$, ** $p<0.01$, *** $p<0.001$)*

Conclusions

This paper reported our efforts to introduce three laboratory modules into one undergraduate-level course (Environmental Engineering Lab and Environmental Engineering) so that students will be exposed to cutting-edge advances in nanotechnology and their impact on the environment.

All modules were fully evaluated with two survey instruments, feedback surveys and multiple choice pre-post-tests. Overall students had positive exposure to nanotechnology and such experience enhanced their learning and increased their interest in science and/or engineering. Modules 1 and 2 are portable and can be easily implemented in other institutions.

Acknowledgement

The authors would like to thank the National Science Foundation (NSF-0532551) and NSF funded Center for High-Rate Nanomanufacturing (NSF-0425826) at the University of Massachusetts Lowell for funding support.

References

1. Wu, C. Sweating the small stuff. *PRISM*, **2004**, 14, http://www.prism-magazine.org.
2. Masciangioli, T.; Zhang, W.X. *Environ. Sci. Technol.* **2003**, March 1, 102A-108A.
3. Zhang, W.X. *J. Nanoparticle Res.* **2003**, *5*, 323-332.
4. Gillham, R.W.; O'Hannesin, S.F. *Ground Water* **1994**, *32*(6), 958-967.
5. Lien, H.L.; Zhang, W.X. *J. Env. Eng.* **1999**, 1042-1047.
6. Lien, H.-L.; Zhang, W.X. *J. Env. Eng.*, **2005**, 4-10.
7. Lien, H.L.; Zhang, W.X. *Colloids and Surfaces A: Physicochem. Eng. Aspects* **2001**, *191*, 97-105.
8. Wang, C.B.; Zhang, W.X. *Environ. Sci. Technol.* **1997**, *31*, 2154-2156.
9. Elliott, D.W.; Zhang, W.X. *Environ. Sci. Technol.* **2001**, *35*, 4922-4926.
10. Naseem, R.; Tahir, S.S. *Water Research* **2001**, *35*, 3982-3986.
11. www.epa.gov/opptintr/lead/index.htm.
12. Kapoor, A.; Viraraghavan, T. *Bioresource Technology* **1995**, *53*, 195-206.
13. Park, K.H.; Park, M.; Jang, H.; Kim, E.; Kim, Y. *Anal. Sci. Technol.* **1999**, *12*, 196-202.
14. Petruzzelli, D.; Pagano, M.; Triavanti, G.; Passino, R. *Solvent Extr. Ion Exchange* **1999**, *17*, 677-694.
15. Srivastav, R.; Gupta, S.; Nigam, K.; Vasudevan, P. *Int. J. Environ. Stud.* **1993**, *45*, 43-50.
16. Pansini, M.; Collella, C. *Desalination* **1990**, *78*, 287-295.
17. Namasivayam, C.; Ranganathan, K. *Water Research* **1995**, *29*, 1737-1744.
18. Hewitt, C.; Metcalfe, P.; Street, R. *Water Research* **1991**, *25*, 91-94.
19. Reed, B.; Arunachalam, S.; Thomas, B. *Environmental Progress* **1994**, *13*, 60-64.
20. Chen, D.; Lewandowski, Z.; Roe, F.; Surapaneni, P. *Biotechnology and Bioengineering* **1993**, *41*, 755-760.
21. Deans, J.R.; Dixon, B.G. *Water Research* **1992**, *26*, 469-472.
22. Seki, H.; Suzuki, A. *J. Colloid Interface Sci.* **1995**, *171*, 490-494.
23. Jang, L.K.; Lopez, B.G.; Eastman, S.L.; Prygogle, P. *Biotechnology and Bioengineering* **1991**, *37*, 266-273.
24. Brown, H.G.; Hensley, C.P. G.L., M.; Robinson, J.L. *Environmental Letter* **1973**, *5*, 103-114.
25. Costley, S.C.; Wallis, F.M. *Water Research* **2001**, *35*, 3715-3723.
26. Veenstra, J.N.; Sanders, D.; Ahn, S. *J. Environ. Eng.* **1999**, *125*, 522-531.
27. Uddin, M.; Chowdhury, A.R. *Integration of nanotechnology into the undergraduate engineering curriculum.* in *International Conferences on Engineering Education.* **2001**. Oslo, Norway.
28. National Nanotechnology Institute, www.nsf.gov/nano
29. Li, F., C., V.; Mohanty, K.K. *Colloids and Surfaces A: Physicochem. Eng. Aspects* **2003**, *223*, 103-112.

Appendix A

Module 1. Synthesis of Palladized Nanoscale Iron Particles

Part I: Preparation of Nanoscale Fe^0 Particles (morning)

(Note: It is important to use fresh palladized nanoscale iron particles in subsequent TCE degradation experiments. Therefore, Modules 1 and 2 are designed as a one day lab exercise.)

Reaction

$$4\ Fe^{3+} + 3\ BH_4^- + 9\ H_2O \rightarrow 4\ Fe^0 \downarrow + 3\ H_2BO_3^- + 12\ H^+ + 6H_2 \uparrow$$

Materials

Sodium borohydride ($NaBH_4$), Ferric chloride ($FeCl_3 \cdot 6H_2O$), 125 mL separatory funnel, 500 and 1000 mL beakers, 500 mL graduate cylinder, Glass stirring rod, Water: Deionized (DI)

Procedure

1. Weigh 4.7290 g sodium borohydride ($NaBH_4$) and dissolve in 500 mL DI water within a 500 mL beaker (0.25M). The sodium borohydride solution is a strong reducing agent. Weigh 6.0759 g ferric chloride ($FeCl_3 \cdot 6H_2O$) and dissolve in 500 mL DI water within a 1-L beaker (0.045M).
2. Fill a 125 mL separatory funnel with the sodium borohydride solution. Position the funnel above the beaker containing the ferric chloride solution.
3. Adjust the separatory funnel stopcock so that the flow comes out as droplets with a speed of about 2 drops/sec.
4. At the same time thoroughly mix the ferric chloride solution with a glass stirring rod.
5. Refill the separatory funnel as necessary until all 500 mL of the sodium borohydride solution has been consumed. The initial color of the ferric chloride solution is rusty, which will go to light green (ferric to ferrous) and eventually to black (ferric to zero valent i.e., Fe^0) as the nano iron crystals are formed.
6. After all the 500 mL of the sodium borohydride solution is consumed, keep mixing for another 10 min.
7. Let the mixture settle for about 45 min and decant the supernatant when the small gas bubbles have almost disappeared. Use the nano iron crystal slurry located at the bottom of the beaker immediately for Module 2.

Part II: Preparation of Nanoscale Pd/Fe Bimetallic Particles (morning)

Reaction

$Fe^0 + Pd^{2+} \rightarrow Fe^{2+} + Pd^0$

Materials

Deoxygenated DI water, 47 mm filter funnel, and 47 mm GF/A filter, Palladium acetate (47.4%), Ethanol, 5 Reaction bottles: i.e., Oakridge Centrifuge Tubes, FEP with screw caps nominal size 50 mL, working volume 45 mL (Agene® No. 3114-0050).

Procedure

Washing of Fe Crystals
1. Transfer the nano crystal iron slurry prepared in Part I (containing an approximate total of 1.25 g nano Fe^0) into a 300 mL beaker.
2. The iron crystals should be washed thoroughly with deionized (DI) water. The total volume of wash water should be at least 600 mL/g of iron. This is an approximate volume of (1.25 g) (600 mL/g) = 750 mL of wash water.
3. Add 250 mL of wash water to the iron crystals in the beaker; mix with a glass stirring rod. Settle for 5 min and decant the clear supernatant.
4. Repeat step 3, two additional times.
5. Add 250 mL of wash water to the iron crystals in the beaker; mix with a glass stirring rod. Transfer 45 mL of well mixed Fe slurry to a reaction bottle. Label the bottle "T0" (i.e., time zero) for use in Part III.
6. Pour the balance of the Fe crystal slurry onto a filter positioned within a filter holder and start the vacuum pump. Wash with 3x (10mL) volumes of DI water. After dewatering, scrape the nano Fe from the filter surface into the palladium acetate/ethanol solution as described in the following steps.

Palletizing of Fe
1. Measure 10 mL ethanol (density = 0.790g/mL) into a 20 mL beaker.
2. Add 0.0012 g (i.e., 1.2 mg) palladium acetate (47.4% Pd) into the ethanol resulting in a 0.015 wt. % palladium acetate/ethanol solution. (Note: For faster TCE degradation add 0.0015 g palladium acetate)
3. Scrape the washed nano Fe^0 from the filter paper into the palladium acetate/ethanol solution within the 20 mL beaker. Thoroughly mix this solution and let the mixture soak for 45 min. This will produce Pd/Fe bimetallic particles.
4. To wash the Pd/Fe bimetallic particles, transfer the contents of the 20 mL beaker to a beaker containing 100 mL of deoxygenated DI water. Mix it well and pour the Pd/Fe bimetallic slurry onto a filter positioned within the filter holder and start the vacuum pump.
5. Scrape the filter into another 100 mL of deoxygenated DI water mix and filter again. Repeat this step 2 times (which results in the Pd/Fe particles

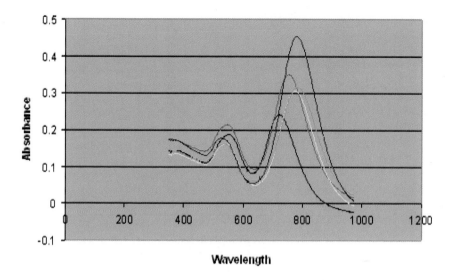

Figure 4.9. Gold Nanorods synthesized with varying amounts of seed and silver nitrate solution. The rods show two peaks, one near 530 nm and one between 600 - 1600 nm. Two surface plasmon bands were observed for the nanorods: the longitudinal plasmon band, corresponding to light absorption and scattering along the particle long axis (second peak); and the transverse plasmon band, corresponding to light absorption and scattering along the particle short axis (first peak).

Color insert - 2

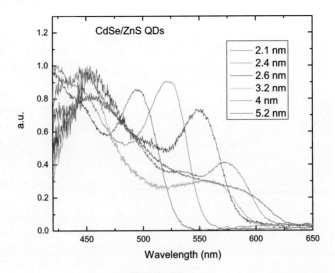

Figure 4.10. Absorption spectra of CdSe/ZnS core-shell QD's with diameters ranging from 1.9 to 5.2 nm. Spectra are normalized. A blueshift in the fundamental absorption edge occurs as the quantum dot diameters reduce from 5.2 to 2.1 nm.

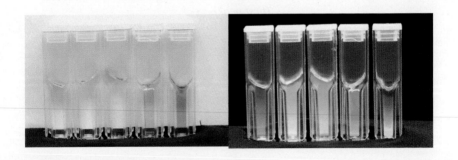

Figure 7.1. Nanometer-sized cadmium selenide sulfide ($CdSe_xS_{1-x}$) and cadmium selenide (CdSe). When the samples are irradiated with ultraviolet light (λ = 365 nm), the emission colors range from violet to orange (right).

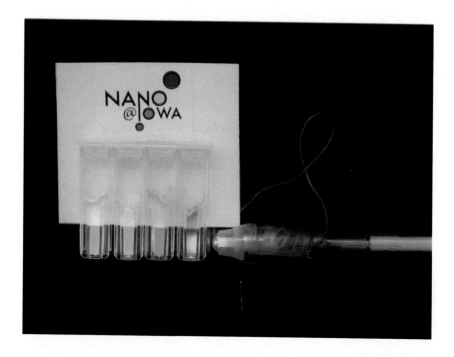

Figure 7.2. Cadmium selenide (CdSe) quantum dot samples embedded in a polymer are used for outreach activities. The luminescent properties can be observed using the attached violet LED pen lights. The NANO@IOWA logo on the cards reinforces the idea that the colors of the quantum dots depend on their size.

Color insert - 4

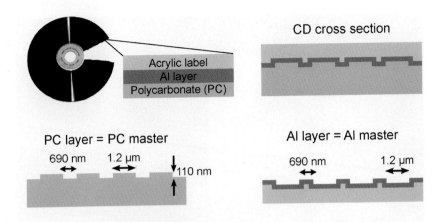

Figure 13.1. Schematic diagram of CD composition

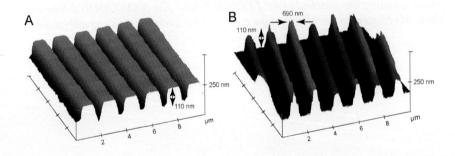

Figure 13.2. AFM images of (A) PC layer and (B) Al layer obtained from CD-R. Either the PC-layer or the Al-layer can be used as a master for soft lithography.

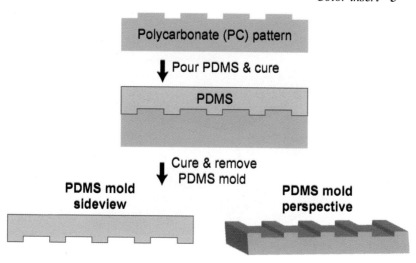

Figure 13.3. Scheme for making PDMS mold using PC layer.

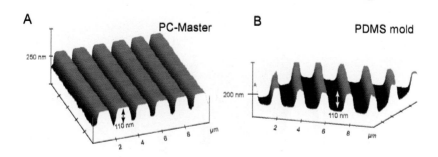

Figure 13.4. AFM image of (A) PC-master and (B) PDMS mold formed from (A).

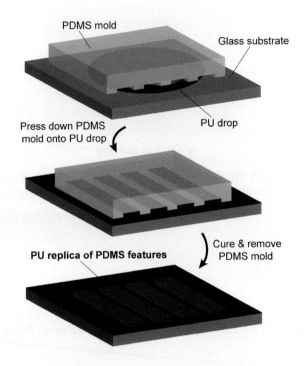

Figure 13.5. Scheme of RM.

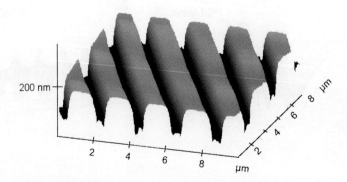

Figure 13.6. AFM image o the PU-replica, which is identical to the PC-master.

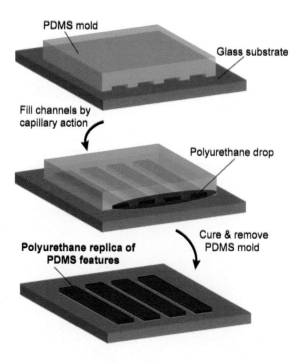

Figure 13.7. Scheme for MIMIC.

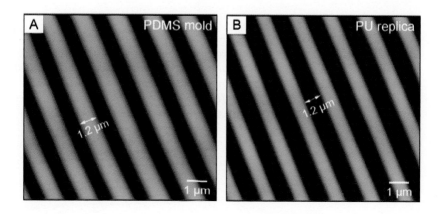

Figure 13.8. AFM images of (A) PDMS mold from Al-layer and (B) PU-replica of the Al-layer after MIMIC. Dark regions correspond to recessed features, and bright regions correspond to raised features.

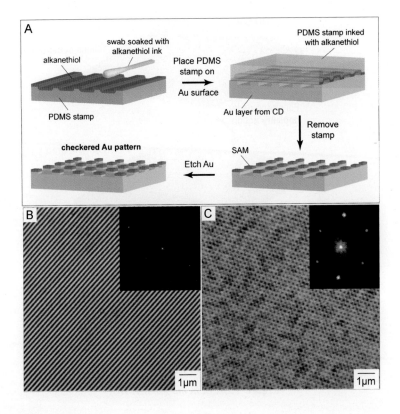

Figure 13.9. (A) Procedure for micro-contact printing and etching of a Au-CD. (B) Optical micrograph of unetched Au-CD and the corresponding linear diffraction pattern (inset). (C) Optical micrograph of the checkerboard pattern of the etched Au-CD and corresponding hexagonal diffraction pattern (inset). In the optical micrographs of (B) and (C), dark regions correspond to recessed features, and bright regions correspond to raised features.

being washed with a total of 300 mL of deoxygenated DI water. After all water has been removed, wash with another 10 mL of DI water.
6. Scrape the washed Pd/Fe bimetallic crystals into a 300 mL beaker. Add 200 mL deoxygenated DI water. Mix it well and evenly transfer to 4 reaction bottles until full. Seal these bottles tightly with the caps.
7. Label the bottles: T1, T2, T3, T4, these will be used in Module 2.

Appendix B

Module 2. TCE Degradation with Palladized Nanoscale Iron Particles (afternoon)

Reaction

$3Fe^0 + C_2HCl_3 + 3H_2O \rightarrow C_2H_4 + 3 Fe^{2+} + 3Cl^- + 3OH^-$

Material

Samples prepared in Module 1 (Part II), 1 sample (T0) of nano Fe^0 (0.25g) within solution in a reaction bottle, 4 samples (T1, T2, T3, T4) of nano bimetallic Fe/Pd (0.25g each) within solution in reaction bottles, TCE, Pentane, Reverse Osmosis (RO) water, 1 Solution storage bottle 2.23 L Reaction bottles: i.e., Oakridge Centrifuge Tubes, 1 Auto pipette (0 – 5 mL), 1 Manual pipette with bulb (4 mL), 10 Dropper pipettes, 20 GC vials 2 mL, 2 Beakers: 200 mL, 20 mL.

Procedures

1. TCE stock solution
   - Put a stir bar into the solution storage bottle. Fill the bottle with RO water completely. It takes 2.23 L water.
   - Use a syringe to inject 80 µL TCE beneath the water surface.
   - Rapidly cap the bottle with a Teflon® tape wrapped rubber stopper. Seal the bottle with Teflon® tape.
   - Store in 20 °C incubator until use (use immediately if possible).
2. TCE/Fe samples
   - Take 2 empty reaction bottles. Fill with RO water. Pipette out and discard 20 mL. Replace with 20 mL TCE stock solution. Label these

samples which represent time-zero control, "CT0", and time-four control, "CT4".
- Take out the reaction bottles (T0, T1, T2, T3, T4) prepared in Part II.
- Pipette out and discard 20 mL supernatant from each bottle and add 3 glass beads to each bottle. Replace with 20 mL TCE stock solution from the storage bottle.
- Put bottle T0 – T4 and CT0 and CT4 on a reaction shaker located within a 20° C incubator for the following (revised) time intervals:

T0	0 min
T1	15 min
T2	30 min
T3	45 min
T4	60 min
CT0	0 min
CT4	60 min

3. TCE extraction and GC samples
   - Hand-shake the reaction bottles T0 and CT0 for 30 seconds. All other samples will be mechanically shaken on the reaction shaker located within the 20 °C incubator for the prescribed time intervals.
   - Following shaking, uncap the bottle and remove 5 mL of mixed solution with an auto pipette, replace with 4 mL of pentane, leaving a small headspace in the tube. Recap the bottle tightly.
   - Put T0 and CT0 bottles on the extraction shaker for 5 min. After 5 min, take them off and centrifuge them for 5 min@ 3000 rpm at 15 °C on the centrifuge (model SL-50T).
   - After removal from the centrifuge, remove the cap and extract 1mL of the pentane located on the top of the water with a pipette and place in a GC vial. Crimp the GC vial immediately.
   - Repeat this step to make a duplicate GC sample.
   - Repeat this extraction procedure for all remaining samples (i.e., T1-T4 and CT4 following the prescribed reaction times.
4. GC Analysis
   - Open the valve of the carrier gas tanks: H_2, Air, N_2.
   - Turn on the power to GC using the switch located on the right side of the GC.
   - Wait for approximately 30 min to equilibrate.
   - Turn on H_2 by turning the button located on the upper left corner of the GC. At the same time, press the ignite button to light the FID.
     - Press signal and check the reading.
     - Wait 1-2 hr until the signal becomes stable (i.e., reading of 0-10).
     - Turn on PC and launch the software for GC signal analysis.
     - The chromatograms will be produced.

5. Typical GC/FID Results Format

Samples		TCE peak area (µV·s)			
Label	Time (min)	Duplicate 1	Duplicate 2	Average	Percentage TCE Remaining
T0	0				
T1	15				
T2	30				
T3	45				
T4	60				
Controls					
CT0	0				
CT4	60				

Appendix C

Module 3. Use of Engineered Nanospheres for Lead Complexation Lab
(This exercise takes one lab session or less assuming nanospheres are available)

Introduction

Lead is a highly toxic metal that was used for many years in products found in and around our homes. Both occupational and environmental exposures to lead remain a serious problem in many developing and industrializing countries, as well as in some developed countries. Lead in wastewater comes mainly from battery manufacturing, printing, painting, dying and other industries.

<u>Toxicity of Lead:</u> Excessive ingestion of lead can cause accumulative poisoning, cancer, nervous system damage, etc. It is a general metabolic poison and enzyme inhibitor and can accumulate in bones, brain, kidney and muscles. Drinking water containing high levels of lead can cause serious disorders, such as anemia, kidney disease and mental retardation. Children 6 years old and under are most at risk, because their bodies are growing quickly. The major route of lead absorption in children is the gastrointestinal tract. The statistics from a recent EPA report shows that in the United States, about 900,000 children ages 1 to 5 have a blood-lead level above the level of concern and children who appear healthy can have dangerous levels of lead in their bodies. The permissible level for lead in drinking water is 15 parts per billion (ppb).

Lead Complexation: Complexation is a conventional technique to remove environmental pollutants. Various complexing compounds e.g. polyacrylic acid (*1*) and caffeic acid (*2*) are in use for lead removal. Lead complexes with these compounds in the form of a coordinate covalent bond. The lead metal ion (or cation) in a complex is called the central atom, the electron-pair donating species from the complexing compound is called the ligand and the number of bonds the central atom can form is called its coordination number.

Functionalized Nanospheres: Carboxyl Functionalized Nanospheres (CFN) synthesized by Dr. Watterson's group at UML have been found to complex free lead ions from aqueous solutions. These functionalized nanospheres were made by a chemoenzymatic method (*3*). First, polyethylene glycols (PEGs) were polymerized to make a long polymer chain, second, hydrophobic groups (long carbon chains) functionalized by carboxyl groups were added to the long polymer chain. These polymers form nanospheres when their concentration is greater than 0.5 g/L in solution. The MW of the CFN used is 12,000 g on average. The complexation reaction can be simply represented by the following reaction:

$$Pb^{2+} + CFN \rightarrow CFN\text{-}Pb^{2+}$$

Objectives

In this lab, the same amount of CFNs will be mixed with various lead concentrations. Final free lead ion concentrations will be measured. The objectives of this lab are:
- determine lead complexation efficiency and
- understand the potential application of nanomaterials for environmental pollutant removal.

Lead Complexation Experimental Procedure

(1) Preparation of lead standard solutions to calibrate the Ion Selective Electrode (ISE): Pipette 1, 3, and 5 mL of 1000 ppm lead perchlorate stock solution into three different 100 mL volumetric flasks and make up to the mark with distilled water to make 10, 30, and 50 mg/L lead standards.

Transfer 25 mL of each lead standard solution into a 100 mL beaker with a 25 mL graduate cylinder. Add 25 mL of Methanol-Formaldehyde solution (with a 25 mL graduated cylinder) and 1 mL of Ionic Strength Adjustor to all the standards. Use these standards to calibrate the ISE. Calibrate the ISE probe according to the procedure on page 4. Now the probe is ready to be used for step 5.

(2) Each group will be given a 100 mL beaker containing a known amount of CFN.

(3) Set up a stir plate and put the beaker on the stir plate. Use speed setting of 3-4 for the stir plate.

(4) Add 25 mL of DI water into the beaker, and put a stir bar into the beaker. Start the stir plate and wait till the CFN is completely dissolved.

(5) Add X mL of 1000 mg/L lead stock solution into the beaker (see Table I). Start the mixing for half an hour (note the starting time and the ending time).

* While waiting, the TA will demonstrate the use of the ISE probe by measuring an initial free lead concentration, i.e., 10 mg/L.

(6) After 30 minutes of mixing, add 25 mL of methanol-formaldehyde solution and 1 mL of ISA solution into the mixture. Measure free lead concentration with the ISE probe and record the value in Table I (with assistance from the TA).

(7) Calculate lead complexation efficiency (use Table I).

Formula

$$\text{Complexation efficiency} = \frac{C_0 - C_e}{C_0} \times 100\%$$

C_0 = initial lead concentration, mg/L
C_e = final equilibrium lead concentration, mg/L

Table I. Lead Complexation Efficiency

	Weight of Nanospheres (g)	Initial free Pb_0^{2+} Concentration (C_0, mg/L)	Lead stock solution needed (mL)	Final free Pb^{2+} Concentration (C_e, mg/L)	Molar ratio (CFN:Pb_0^{2+})	Efficiency (%)
Group 1	0.0435	6	0.15			
Group 2	0.0435	10	0.25			
Group 3	0.0435	15	0.375			
Group 4	0.0435	30	0.75			

References

1. Morlay, C.; Mouginot, Y.; Cromer, M.; Vittori, O. *Can. J. Chem.*, **2001**, *79*(4), 370–376.

2. Boilet, L., Cornard, J.P. and Lapouge, C. *J. Phys. Chem. A*, **2005**, *109*, 1952 - 1960.

3. Kumar, R.; Chen, M.H.; Parmar, V.S.; Samuelson, L.A.; Kumar, J.; Nicolosi, R.; Yoganathan, S.; Watterson, A.C. *J. Am. Chem. Soc.,* **2004***, 126*, 10640-10644.

Appendix D

Feedback Survey

Evaluation of Nanotechnology Modules
University of Massachusetts, Lowell - Spring 2007

Please answer the following questions to help us learn more about the students who are enrolled in this course and to evaluate and improve the quality of modules taught in this course. All responses will remain strictly confidential.

Course Name: _____

Why did you enroll in this course?: ○ Requirement   ○ Elective

Year in School: ○ Freshman   ○ Sophomore   ○ Junior   ○ Senior   ○ Other _____

Sex: ○ Female   ○ Male

Ethnicity (please select all that apply):
○ African American/Black   ○ Asian   ○ Caucasian/White   ○ Hispanic/Latino(a)
○ Native American   ○ Pacific Islander   ○ Other (please specify) _____

Do you plan to attend graduate school?   ○ Yes   ○ No   ○ Don't Know

Please respond to each of the following statements by filling in the bubble that best reflects your opinion.

	Strongly Disagree	Disagree	Neither Agree nor Disagree	Agree	Strongly Agree
I have a better understanding of nanotechnology after completing this module.	○	○	○	○	○
I would like to experience additional nanotechnology modules in other courses.	○	○	○	○	○
My interest in science (and/or engineering) has increased as a result of this nanotechnology module.	○	○	○	○	○
I am interested in conducting further research related to nanotechnology.	○	○	○	○	○

Please answer the following questions. If you need additional space, please use the back of this survey.

1. What did you like about this module? _____

2. What was most challenging about this module? _____

3. Any other comments or suggestions that would improve the quality of this module and/or your learning experience? _____

THANK YOU FOR COMPLETING THIS SURVEY!

Appendix E

Synthesis of Palladized Nanoscale Iron Particles

Multiple Choice Questions

1) What role does nanoscale iron play in groundwater treatment?
   a) It kills bacteria
   b) It dechlorinates organic contaminants
   c) It produces a barrier to ground water flow
   d) All of the above
2) In the synthesis of nano iron, iron is reduced from iron (III) to elemental iron. What is used as the reducing agent?
   a) Elemental hydrogen
   b) Chloride ion
   c) Hydride ion
   d) Borohydride ion
3) What gas is evolved during the synthesis of nano irons?
   a) Hydrogen
   b) Chlorine
   c) Oxygen
   d) None of the above
4) What reaction conditions are used to grow uniform nano-sized particles?
   a) A large excess of reducing agent
   b) Proper stirring
   c) Controlled rate of addition of reagents
   d) All of the above
5) Why are the nanoscale Pd/Fe bimetallic particles stored in deoxygenated water rather than drying in air?
   a) The particles will oxidize in air and become deactivated
   b) The particles are toxic when dry
   c) If dried the particles become too reactive to use in groundwater remediation
   d) All of the above
6) What is the advantage of using nanoscale iron in comparison to commercially available powdered zero-valent iron ?
   a) Nanoscale Fe^0 has a higher surface area
   b) Nanoscale iron can be injected into an aquifer
   c) Nanoscale Fe^0 has a higher reaction rate
   d) All of the above
7) Why are the nanoscale Fe^0 particles coated with palladium?
   a) To preserve the particles
   b) Palladium acts as a catalyst
   c) To decrease the reactivity of the particles
   d) None of the above

8) What is the typical maximum size of nanoscale particles used in groundwater treatment?
   a) 1 nm
   b) 1000 nm
   c) 10 nm
   d) 100 nm

Appendix F

Module 2 Pre- Post-Test
TCE Degradation with Palladized Nanoscale Iron Particles
Multiple Choice Questions

1) Which of the following statements are true with respect to first-order reactions?
   a) The rate constant (k) is independent of the remaining contaminant concentration
   b) The rate of contaminant decay (mass/time) does depend on the concentration of contaminant remaining
   c) The time required for a given fraction of a contaminant to decompose is independent of the initial concentration
   d) All of the above

2) Why do we use pseudo first-order kinetics to characterize the reaction of TCE and Pd/Fe bimetallic particles?
   a) It is a well documented first-order reaction
   b) It is used as a first approximation when variables are not well understood
   c) Zero-order analysis requires extensive analysis and does not improve fit
   d) All of the above

3) What are the byproducts of <u>TCE contaminated ground water treatment</u> with nano irons?
   a) Hydrocarbons and elemental chlorine
   b) Hydrocarbons and elemental hydrogen
   c) Hydrocarbons and chloride ions
   d) Hydrocarbons and carbon dioxide

4) Why do we add pentane to the reaction bottle at the end of the kinetic experiment?
   a) To serve as a reaction catalyst
   b) To inhibit by-product formation
   c) To extract TCE for GC measurement
   d) To minimize the risk of TCE exposure

5) If the rate constant (k) is known, which of the following statements are true?
   a) It is possible to predict the composition of the degradation byproducts.
   b) The contaminant half-life can be predicted

c) The original contaminant concentration can be calculated
   d) All of the above
6) Kinetic analysis to determine a first-order rate constant requires which of the following?
   a) Statistical analysis of data
   b) Use of the quadratic equation
   c) Use of geometry
   d) Graphical analysis of data

Appendix G

Module 3 Pre- Post-Test

Use of Engineered Nanospheres for Lead Complexation Lab
Multiple Choice Questions

1) What are the toxic effects of Lead in drinking water?
   a) mental retardation
   b) anemia
   c) cancer and accumulative poisoning
   d) all of the above
2) What is the EPA limit for lead concentration in drinking water?
   a) 1 mg/L
   b) 100 ppb
   c) 15 ppb
   d) no limit
3) What could be done in order to increase the complexation efficiency of the nanospheres used?
   a) Increase the amount of nanospheres used
   b) Decrease the amount of nanospheres used
   c) Increase surface area.
   d) Decrease surface area.
4) What is the function of the Ion Selective Electrode used in this lab?
   a) detect nanosphere concentration
   b) detect the concentration of free lead ions in aqueous solutions
   c) detect ions other than lead in solution
   d) detect the concentration of nanosphere and lead solution mixture
5) How is lead complexation efficiency calculated?
   a) initial concentration – final concentration
   b) final concentration/initial concentration
   c) (initial concentration – final concentration)/100
   d) [(initial concentration – final concentration)/initial concentration]*100

Chapter 9

Scanning Tunneling Microscopy and Single Molecule Conductance

Novel Undergraduate Laboratory Experiments

Erin V. Iski[1], Mahnaz El-Kouedi[2], and E. Charles H. Sykes[1,*]

[1]Department of Chemistry, Tufts University, Medford, MA 02155
[2]Department of Chemistry, University of North Carolina at Charlotte, Charlotte, NC 28223

> An undergraduate physical chemistry laboratory has been developed in which students study a coadsorbed self-assembled monolayer (SAM) system on Au(111). The experiment was adapted from the work of Bumm et al. and modified to fit into an undergraduate physical chemistry laboratory setting. The system was studied with a teaching Scanning Tunneling Microscope (STM). Through this laboratory exercise, students will learn experimental techniques such as the use of scanning probes, and quantum mechanical concepts like electron tunneling and molecular conductance. Students will also make the connection between the study of SAMs in the laboratory and the commercial use of SAMs in nanotechnology today.

Introduction

Self-assembled monolayers (SAMs) have been central to the development of the field of nanotechnology. SAMs are composed of a head group, usually sulfur, which forms a strong covalent bond with a gold surface, and an alkyl chain that may or may not contain other functional groups (1). The length of the alkyl chains can range from 1 to over 30 CH_2 units in length. Due to both the

© 2009 American Chemical Society

interaction between the sulfur head groups with the substrate and the van der Waals interactions between the alkyl tail groups, SAMs are very well ordered on the molecular scale. Their ease of preparation and stability allow for application in both research and industry (*2,3*). For example, SAMs are used as the matrix into which the molecules used for molecular switches are embedded (*4*). They are also used to tether DNA to Au nanoparticles, which can subsequently assemble and change color for colorimetric detection applications (*5*). SAMs have found applications in sensor technologies where their self-assembly and functional diversity can be exploited (*6*). They have also been used to create "molecular rulers" in which the length of the SAM molecules can be manipulated in order to precisely dictate the width of nanowires (*7*). Finally, SAMs have been applied to microelectromechanical systems (MEMS) in order to reduce friction, since traditional oil-based lubricants are not desirable due to viscosity problems which can overwhelm the functioning of the systems themselves (*8*).

Molecular conductance through SAM chains is of particular interest in the field of nanoscale molecular electronics (*4,9,10*). Better understanding and quantification of electron transfer within these molecular frameworks will enable more unique and novel applications in the future (*11,12*). Scanning tunneling microscopy (STM) is well suited for these types of measurements because it is based on electron tunneling which is very sensitive to small changes in distance between the STM tip and surface, allowing STM to measure very small surface features including atoms and adsorbed molecules. An example of STM images of thiol SAMs are shown in Figure 1.

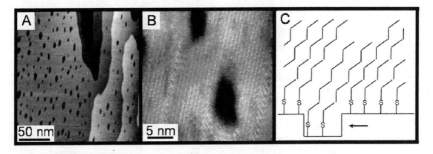

Figure 1. A) A 200 x 200 nm² STM image of a SAM of C_{10} thiol. The dark holes in this image are called "etch pits". These pits are areas where one layer of Au atoms is missing. B) A 5 x 5 nm² image of a C_8 thiol SAM. Each small bright spot is an individual C_8 thiol molecule. C) Schematic showing the ordering of a self-assembled monolayer of C_8 thiol. The lower area highlighted by the arrow represents an etch pit as seen in Figure 1A.

This lab focuses on both imaging SAMs with STM and making measurements that reveal both their physical and electronic properties.

All of the fundamental experiments for this laboratory exercise were performed on a high-end research STM by Bumm et al. at Penn State (*13*). The contribution of this paper is the adaptation of that work to an undergraduate laboratory setting. Cost effective teaching microscopes are now on the market, which make experiments such as this accessible to a wider variety of students. Teaching STMs have the capability to atomically image substrates, such as highly oriented pyrolytic graphite (HOPG), and to obtain molecular resolution of short-chained thiol molecules. A low-current amplifier that can measure currents as low as 0.02 nA is necessary to image SAMs, but is readily available on such teaching instruments. It was necessary to reduce the length of the alkyl chains from those used in the original experiment due to the current limitations of a teaching STM. In the original paper a C_{12} thiol was used, but this molecule is too insulating to be imaged with a teaching STM.

For the undergraduate physical chemistry lab, students are introduced to STM instrumentation including STM components, STM operation, cutting of tips, scanning and analysis of data, as well as the theory behind electron tunneling and molecular conductance. The experiment is also set up so that students not only image a two-component coadsorbed SAM, but they also use the measured STM height difference to determine the chain length of one of the thiol molecules. The identity of one of the thiol molecules comprising the two-component SAM is known, while the identity of the second component is determined upon completion of the lab. In order to determine the identity of the unknown thiol molecule, students must understand that the height as measured by the STM is not the same as the actual physical height of the molecule; students are thus introduced to the idea that electronic as well as topographic information is contained in STM images. In addition, the experiment strengthens the student's skills with Excel® spreadsheets, statistics, and data analysis, while introducing them to a system that is very important to modern day nanotechnology.

Scanning Tunneling Microscopy

The typical layout of a STM, which was invented in 1981 by Binnig and Rohrer at IBM, is shown in Figure 2 below (*14*).

In STM, a sharp metal tip, most commonly made from a platinum/iridium or tungsten wire, is brought within a nanometer of a conductive surface (*15*). STM tips are made atomically sharp by either the physical act of cutting the end of the metal wire or through an electrochemical etching procedure. Following classical mechanics, no current can flow between the tip and the surface as they are not in metallic contact. However, at tip-surface separations on the order of nanometers, the wavefunctions of the atoms in the surface and the single atom at the end of the tip overlap, and electrons can quantum mechanically *tunnel* from the tip to the surface or vice versa when a voltage is applied (*16*). It is this extreme dependence of the tunneling current (*I*) on the tip-surface separation that allows STM to achieve atomic-scale resolution, i.e. a very small change in surface-tip separation leads to a large change in tunneling current. The equation for the tunneling current is as follows: $I \propto \exp(-2\kappa d)$, where *d* is the distance between the surface and tip and

$$\kappa^2 = \frac{2m}{\hbar^2}(eV_B - E) \qquad (1)$$

where $\hbar$ is Planck's constant ($h/2\pi$), E is the energy of the state from which tunneling occurs, and eV_B is the barrier height.

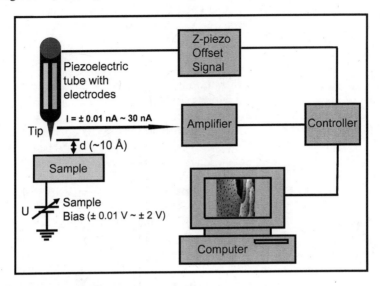

Figure 2. Schematic of a STM. The atomically sharp tip is mounted in a piezoelectric tube-scanner and is controlled by a feedback loop, which ultimately produces an image on the computer screen.

The atomically sharp STM tip is mounted on a piezoelectric tube scanner. When a voltage is applied across piezoelectric materials, they expand or contract. The piezoelectric tube scanner used in the STM is often built in three parts so that the application of voltages to each part individually gives the possibility of very accurate movement in the x, y, and z directions (*16*).

An enhanced explanation of electron tunneling involves an understanding of the Fermi energies (the energy of the highest occupied electronic state) of the tip and the surface (*7*). When the STM tip and the surface are brought close together, but are not in metallic contact, the offset in the Fermi energies of the sample, E_F^S, and the tip, E_F^T, is equal to the difference in the work functions ($\Delta\phi$) of the two metals, $\Delta\phi = E_S^F - E_T^F$. The work function is described as the minimum energy needed to remove an electron from the Fermi level to the "vacuum level". The "vacuum level" is the energy level associated with a stationary electron that is an infinite distance away from the surface.

When the sample is biased positively with respect to the STM tip, the Fermi energy of the sample is lowered by eV, where e is the elemental charge and V is the potential difference. The Fermi level of the tip now lies at the same energy as some of the unoccupied states of the sample, and thus it is energetically favorable for electrons to tunnel from the tip to the sample since the unoccupied states of the sample are at a lower energy (*15*). If the sample is instead biased negatively with the respect to the STM tip, the tunneling will occur in the opposite direction.

STM imaging is most often performed in one of two modes of operation. In the first mode, called the *constant height mode*, the tip is scanned across the *xy*-plane of the surface without any changes of the *z*-piezo (*16*). Due to the fact that the tunneling current is exponentially dependent on the distance between the tip and the surface, variations in the tunneling current can be recorded and correlated to changes in the topography of the surface. While images can be acquired very quickly in this mode, it is rarely used, as scanning over any surface features larger than the tip-sample distance will "crash" the atomically sharp STM tip into the surface. In the second mode of operation, called the *constant current mode*, the tunneling current is kept constant by a feedback loop that continually adjusts the tip height in the *z*-direction. An image is generated by plotting the changes in the *z*-piezo height versus the *xy*-position of the tip and contains information about both sample topography and electronic structure.

Imaging of SAMs with STM

The fundamental concepts expressed in this portion of the paper are derived from the work performed by Bumm et al. (*13*). The height of a film as measured by the STM, h_{STM}, is often not the same as the actual physical height of the film, h_{film}, since the tip-sample interface combines both physical and electronic properties of the surface (*13*). When imaging a SAM, the STM tip hovers just above the film (SAM), meaning that the STM tip does not penetrate into the monolayer, thus creating two distinct regions: the vacuum gap (d_{gap}) and the actual film or SAM (h_{film}). The height of the film as measured by the STM is thus described by the following equation,

$$h_{STM} = h_{film} + d_{gap} \qquad (2)$$

The transconductance (reciprocal resistance) of the tip-film gap ($G_{gap} = 1/R$) is adjustable and depends on the separation between the tip and the SAM, d_{gap}. This separation is controlled by the STM. While scanning in the constant-current mode, the imaged topography includes height contributions from both the film and the tip-film gap since the tip does not touch the top of the SAM (*13*).

In this lab, a two-component SAM comprised of regions of an octanethiol (C_8) SAM intermixed in regions of a C_x SAM, which is longer than the C_8 component, will be used for measurements. The identity of the C_x SAM for the development of this experiment was decanethiol (C_{10}). The transconductance ($G = 1/R = I/V$) of a molecule decreases exponentially with increasing length, as increasing length results in the molecule becoming more insulating (*13*). As the STM scans from a region of the SAM that contains alkyl chains with eight carbons to a region in which the chains contain ten carbons, h_{film} increases by a distance of 2 CH_2 units (Each CH_2 unit is 1.1 Å in length.). Since the longer chain is more insulating, the measured transconductance or current decreases. In order to compensate for this decrease in measured tunneling current, the STM decreases the d_{gap} by 1.1 Å to increase the tip-film gap transconductance (G_{gap}), which allows for a constant overall transconductance. This is explained visually in Figure 3.

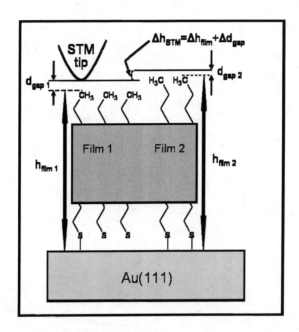

Figure 3. Schematic showing the STM tip path over a multi-component SAM. To maintain a constant overall current, changes are made in the tip-film gap, d_{gap}, to compensate for differences in chain lengths (13).

The relationship between the height as measured by the STM and the actual height difference between two SAM regions with different alkyl chain lengths can also be explained through various quantum mechanical equations (*13*). As stated previously, the tunneling current used by the STM to image is exponentially related to the tip-substrate separation. The tunneling current is related to resistance according to Ohm's Law ($V = IR$) and transconductance (G) is reciprocal resistance ($G = 1/R$). Thus, it can be derived that G is also exponentially related to the tip-substrate separation. The tip-substrate separation can be broken down into two portions as represented in Figure 3 (*13*). The transconductance of these two regions can then be described as:

$$G_{gap} = A e^{-\alpha d_{gap}} \tag{3}$$

$$G_{film} = B e^{-\beta h_{film}} \tag{4}$$

where α and β are decay constants and A and B are contact conductances. The transconductance of the system is the product of the two individual transconductances: $G_{tot} = G_{film} \times G_{gap}$.

Through some algebraic manipulation and approximation, the following equation can be derived, which relates the height of the film as measured by the STM to the actual physical height of the film (For a complete derivation of the equation, see ref. 13.).

$$\Delta h_{STM} = \Delta h_{film}(1 - \beta/\alpha) \qquad (7)$$

According to theory, the values for α and β are 1.2 Å^{-1} and 2.3 Å^{-1} (*17*), respectively. It is therefore possible to write:

$$\Delta h_{STM} = 0.5 \Delta h_{film} \qquad (8)$$

Thus, a mathematical relationship between the measured height differences (Δh_{STM}) and the actual height differences (Δh_{film}) of a multi-component SAM can be obtained (*13*).

Step by Step Procedure

1) Preparation of surface:

   a) The Au(111) surfaces, prepared by vapor deposition of Au onto clean mica, can be purchased from Agilent Technologies. The SAM molecules can be purchased from Aldrich Chemical Co. Prior to SAM formation, the Au(111) surface is annealed in a hydrogen flame. For a full explanation of how to flame anneal Au/mica, see the Agilent Technologies website (http://www.home.agilent.com).

   b) The clean Au substrate is then placed into a 1 mM solution of octanethiol (C_8) in ethanol. The C_8 SAM is allowed to form for 12 hours, after which the sample is removed from solution, washed with ethanol, and dried with He (g).

   c) The sample is next placed into a vial containing a single drop of neat decanethiol (C_{10}) that is not in direct contact with the sample, and the vial is heated to 78 °C for 2 hours. The sample is then removed, rinsed in ethanol, and blown dry with He (g) (*18*). This procedure produces a mixed SAM of C_8 and C_{10} thiol molecules. Due to the amount of time it takes to create these samples, they are generally prepared for the students prior to the laboratory session.

2) Mounting the sample on the sample plate:

   a) After SAM formation, the sample is attached to the metal sample plate using double-sided tape. The sample is then electrically connected to the sample plate using conductive silver paint.

3) Scanning the sample:
   a) The metal sample plate is next placed on the magnetic end of the sample holder, which can then be inserted into the STM.
   b) The STM tips are made from Pt/Ir wire cut with pliers to create an atomically sharp tip, and all images are acquired in the constant current mode. All of the images in this paper were obtained on a Nanosurf easyScan™ teaching STM with a low-current pre-amplifier. For a more detailed procedure, refer to http://ase.tufts.edu/chemistry/sykes/research/samlab.pdf.

Data from the Experiment

Figure 4 demonstrates the type of STM images that students obtained during this laboratory exercise. The darker regions of the surface indicate areas which are covered with the C_8 thiol SAM. The brighter regions in the images signify regions of the longer C_{10} thiol SAM. Etch pits are also observed.

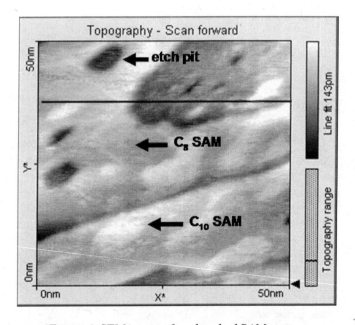

Figure 4. STM image of coabsorbed SAM system.

From these images students take height measurements in order to investigate the height changes between the different SAM domains on the surface. Students use the known height of a Au step (0.235 nm) to calibrate their measurements. They correlate the measured height changes with the theoretical height difference between two different SAM chains. Through basic conductance equations they can understand why the height difference as measured by the STM is not the same as the actual physical height difference between the two different SAM chains. They also determined the identity of the longer C_x chain. A line scan measuring the height differences across the surface along the area marked by the black line (see Figure 4) is shown in Figure 5. Students took their measurements from these line scans.

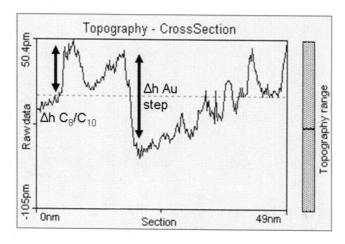

Figure 5. A line scan measuring the height changes across an area of the surface.

The entire experiment fits well within a three-hour lab period. On average the students were able to cut atomically sharp, usable tips within three attempts. The students were able to achieve molecular resolution within 30 minutes of starting the experiment. The students used an Excel spreadsheet to keep track of their data, which also allowed for easy mathematical manipulation in terms of calibration and averaging. From the students who have attempted this experiment, the feedback has been quite positive. Most students were intrigued with the hands-on use of a STM, which is typically reserved for research purposes. The students also stated that the experiment helped to solidify the somewhat abstract, scientific concepts of electron tunneling and molecular conductance.

The students were asked the following questions after completing the experiment.

Post-Lab Questions

1. What is the topographic height difference between a C_8/C_x SAM as measured by the STM?
2. What is the expected height difference between a C_8/C_x SAM based on the actual physical height difference of the molecules?
3. How can the difference between these values be explained?
4. How many carbons are in the C_x SAM?

Fundamental scientific questions

1. How is it possible for electrons to move from the tip to the substrate if they are not in physical contact?
2. Why does the tip have to be atomically sharp?
3. What other systems can be imaged using STM?

Conclusions

Teaching STMs can obtain molecular resolution of SAMs of appropriate length; chains containing more than 12 carbons are difficult to resolve with teaching STMs, as they become too insulating, making the observation of a measurable current difficult. The complexity of the laboratory experiment can vary from basic imaging for an instrumental lab to the complex physics behind electron tunneling and electron conductance through molecules for an advanced physical chemistry lab. Through this exercise, students will have the chance to learn about scanning probes, molecular packing structures, SAMs, electron tunneling, and molecular conductance. They will also have the opportunity to learn about an important aspect of nanotechnology, which allows them to connect with current advances in this exciting and new area of science.

Future Applications

In the future it may be possible to expand this lab to incorporate more recent advances with SAMs in nanotechnology, such as nanoparticle adsorption on SAMs (*19*). It may also be possible to include current vs. voltage (IV) spectroscopy with the teaching STM on samples with and without SAMs, which provides a local probe of electronic states. IV curves can be used to determine whether or not an organic film is present on a surface, and therefore, in a sense, provide a small degree of chemical identification.

References

1. Laibinis, P. E.; Whitesides, G. M.; Allara, D. L.; Tao, Y. T.; Parikh, A. N.; Nuzzo, R. G. *J. Am. Chem. Soc.* **1991**, *113*, 7152-7167.
2. Poirier, G. E. *Chem. Rev.* **1997**, *97*, 1117-1127.
3. Stranick, S. J.; Parikh, A. N.; Tao, Y. T.; Allara, D. L.; Weiss, P. S. *J. Phys. Chem.* **1994**, *98*, 7636-7646.
4. Donhauser, Z. J.; Mantooth, B. A.; Kelley, K. F.; Bumm, L. A.; Monnell, J. D.; Stapleton, J. J.; Price, D. W.; Rawlett, A. M., Allara, D. L.; Tour, J. M.; Weiss, P. S. *Science* **2001**, *292*, 2303-2307.

5. Elghanian, R.; Storhoff, J. J.; Mucic, R. C.; Letsinger, R. L.; Mirkin, C. A. *Science* **1997**, *277*, 1078-1081.
6. Flink, S.; van Veggel, F. C. J. M.; Reinhoudt, D. N. *Adv. Mater.* **2000**, *12*, 1315-1328.
7. Hatzor, A.; Weiss, P. S. *Science* **2001**, *291*, 1019-1020.
8. Li, S.; Cao, P.; Colorado, R.; Yan; X.; Wenzl, I.; Shmakova; O. E.; Graupe, M.; Lee, T. R.; Perry, S. S. *Langmuir* **2005**, *21*, 933-936.
9. Akkerman, H. B.; Blom, P. W. M.; de Leeuw, D. M.; de Boer, B. *Nature* **2006**, *441*, 69-72.
10. Lindsey, J. S. *New J. Chem.* **1991**, *15*, 153-180.
11. Kushmerick, J. J.; Pollack, S. K.; Yang, J. C.; Naciri, J.; Holt, D. B.; Ratner, M. A.; Shashidhar, R. *Ann. N.Y. Acad. Sci.* **2003**, *1006*, 277-290.
12. Heath, J. R.; Ratner, M. A. *Phys. Today* **2003**, *56*, 43-49.
13. Bumm, L. A.; Arnold, J. J.; Dunbar, T. D.; Allara D.L.; Weiss, P. S. *J. Phys. Chem. B* **1999**, *103*, 8122-8127.
14. Binning, G.; Rohrer, H.; Gerber, C.; Weibel, E. *Phys. Rev. Lett.* **1982**, *49*, 57-61.
15. Kolasinski, K. W. *Surface Science: Foundations of Catalysis and Nanoscience*; John Wiley & Sons: West Sussex, England, **2004**.
16. Attard, G.; Barnes, C. *Surfaces*; Oxford University Press: New York City, **2004**.
17. Salmeron, M.; Neubauer, G.; Folch, A.; Tomitori, M.; Ogletree, D. F.; Sautet, P. *Langmuir* **1993**, *9*, 3600-3611.
18. Donhauser, Z. J.; Price, D. W.; Tour, J. M.; Weiss, P. S. *J. Am. Chem. Soc.* **2003**, *125*, 11462-11463.
19. Yang, G.; Tan, L.; Yang, Y.; Chen, S.; Liu, G. *Surf. Sci.* **2005**, *589*, 129-138.

Chapter 10

Nanotechnology for Freshmen

A Review of Nanomaterial Synthesis Experiments for the General Chemistry Laboratory Course

Kurt Winkelmann

Department of Chemistry, Florida Institute of Technology, Melbourne, FL 32901

This chapter provides an overview of nanotechnology-themed experiments for general chemistry laboratory instructors. Prepartions of CdS, Fe_3O_4, Ag and Au nanoparticles and Ni nanowires are presented as examples. Each experiment's description includes theoretical background, procedures and applications of the nanomaterial. Additional resources for teaching nanotechnology to freshmen and designing new experiments are provided.

There is no no reason why general chemistry students cannot synthesize the same advanced materials that are part of satellite gyroscopes, solar energy cells and DNA diagnostic tests. All these applications, and many others, use nanomaterials that students can easily make and study. Scientists have a significant interest in nanotechnology due to the changes in physical and chemical properties that occur as particle size decreases to less than one micron. A nanomaterial's interaction with light and its surface properties are among the many characteristics that can greatly differ from the analogous bulk particles. Educators are exploring ways of incorporating this cutting edge field of research into their courses in order to spark student interest in science, math and engineering (*1-4*) and to prepare students for the many nanotechnology-based careers expected in the future global economy (*3, 5*). Nanotechnology topics provide an opportunity for students to learn how new scientific discoveries influence society (*6*).

How can general chemistry students learn about nanotechnology, a field that is based on quantum mechanics and materials science? Obviously, instructors must simplify the theory and usually provide only a qualitative explanation. Despite the lack of detail, these experiments offer a unique opportunity for first-year college students to learn about nanotechnology. Many ideas for relating nanotechnology to freshmen and even high school students can be found in chemistry education journals and web sites (see Resources). The field of nanotechnology is interdisciplinary. This shows students that knowledge of chemistry is necessary to understand important concepts in other fields of science and engineering. For instance, these experiments illustrate that nanoparticles (also known as quantum dots) are inherently unstable due to their high ratio of surface atoms to interior atoms. Nanoparticles spontaneously aggregate to form bulk particles. Topics covered in a typical general chemistry course, such as intermolecular forces, can be used to devise strategies that prevent this from occurring.

This chapter provides an overview of the state of nanomaterial synthesis in the general chemistry laboratory today. Each experiment's description includes the theoretical background, procedure and applications for each nanomaterial. Useful resources for further investigation are provided at the end of the chapter. It is hoped that this information will help general chemistry instructors implement nanomaterial synthesis experiments in their laboratory curriculum and inspire them to develop new nanotechnology experiments.

Nanomaterials presented in this chapter (CdS, Fe_3O_4, Ag, and Au nanoparticles, and Ni nanowires) were selected because they are simple to prepare, highlight the broad range of experimental techniques used, and have many interesting properties and applications. This is not an exhaustive list of experiments for first-year students and there are also many more challenging synthesis and characterization experiments designed for advanced lab courses. The instructor should provide students with background reading and give a short introductory lecture that explains any unfamiliar nanotechnology concepts. Working in pairs, students should complete each experiment within a three hour period.

All of the experiments described in this chapter have been performed by freshmen students in Florida Tech's Nanoscience and Technology (NST) introductory laboratory course. Since the only prerequisite of the NST lab is completion of one semester of general chemistry, implementing these experiments in a general chemistry laboratory should not be difficult. The experiments are designed to use common reagents and laboratory equipment. Procedures yield only a small amount of solution containing nanoparticles so that the volume of chemical waste is minimal.

Some nanoscale properties are more effectively demonstrated with the use of a UV/visible spectrophotometer but none of these experiments is completely reliant on it. More advanced instrumentation, such as scanning probe and electron microscopes, could be used to characterize students' samples. Such characterization methods might be added in order to make these experiments appropriate for physical chemistry or instrumental analysis lab courses.

Nanomaterials may pose some risk to humans and the environment. The extent of the dangers is hard to quantify and much more research needs to be done in this area (*7, 8*). Experiments described in this chapter yield nanosized products dispersed in liquid media. This greatly limits the exposure of students to accidental inhalation of these materials. Goggles and lab coat or apron are strongly recommended as a minimum amount of protection. When appropriate, the experimental procedure lists other precautions. All reagents should be disposed of properly according to the appropriate state and federal laws. Given the unknown impact of nanomaterials in the environment, it is prudent to allow the nanomaterials to agglomerate and then dispose of the bulk material.

Cadmium Sulfide Nanoparticles

Synthesis of CdS nanoparticles can be performed easily and safely by freshmen students. Based originally on research by Agostiano (*9*), the procedure has been adapted to use reagents commonly available to a general chemistry laboratory (*10, 11*). This experiment illustrates how intermolecular forces affect the formation of micelles and how surfactants behave in oil-water mixtures. The difference in color between the bulk and nanosized CdS is visibly obvious but students can also calculate the nanoparticle size with the aid of a UV/visible spectrophotometer. The explanation for the color difference is based on quantum confinement of electrons and holes in the particle's semiconductor lattice.

Students should be familiar with the reaction that forms the CdS nanoparticles shown in equation 1:

$$Cd^{2+}_{(aq)} + S^{2-}_{(aq)} \rightarrow CdS_{(s)} \qquad (1)$$

Such a reaction is often used to illustrate precipitation reactions and is part of general chemistry qualitative analysis schemes. In this experiment, students watch that same reaction occur in a nanoscale drop of solution that limits the availability of reactant ions. Students should review the relationship between wavelength, energy and color (absorbed and reflected) to help them understand why nanosized particles have a different color than bulk CdS.

Background

Of the many properties that change on the nanoscale, color is the easiest to observe. Bulk cadmium sulfide crystals are orange because the material absorbs light below 515 nm. This wavelength corresponds to a band gap energy of 2.4 eV (wavelength in nm × energy in eV = 1240 nm eV), indicating that CdS is a semiconductor. Upon absorption of a sufficiently energetic photon, an electron in the crystal lattice is excited from its ground state (the valence band) to an excited state (the conduction band). The excited electron can be thought of as occupying an orbital that extends over the whole crystal, so it is no longer localized in one valence band lattice position. An absence of an electron in the valence band is referred to as a hole (h^+_{vb}). As a nearby electron moves to fill in the position formerly occupied by the excited electron, the positive charge moves in the opposite direction. This leads to the concept that the hole behaves like a particle which is delocalized within the valence band.

Electrons and holes move through the crystal and are attracted to each other. Together they form a hydrogen atom-like system called an exciton. Since the electron and hole must remain within the particle or on its surface, the exciton's radius can be constrained if the CdS particle's diameter is small. Analogous to the simple particle in a box model, confining the exciton to a smaller region raises the energy difference between energy levels. Therefore, as the nanoparticle's size decreases, the semiconductor's band gap energy increases. This in turn changes the optical absorption spectrum and the color of the material (12). The wavelength of light required to excite an electron blue shifts, causing the color to change from orange to yellow to eventually a green-yellow color at extremely small CdS particle sizes.

The effective mass model can be used in conjunction with the UV/visible spectrum of the CdS colloid to determine the approximate particle size (13). In this model, the "effective mass" of electrons and holes is determined by their interactions with other electrostatic charges within the lattice. Equation 2 shows how the semiconductor band gap is related to the particle size:

$$E_g^{nano} \cong E_g^{bulk} + \frac{h^2}{8r^2}\left(\frac{1}{m_e^*}+\frac{1}{m_h^*}\right) - \frac{1.8e^2}{4\pi\varepsilon\varepsilon_o r} \qquad (2)$$

Equation 2 uses the following values for variables and physical constants:

- E_g^{nano} = band gap energy (in joules) of the nanoparticle as determined from the UV/visible absorbance spectrum (see below),
- E_g^{bulk} = band gap energy of bulk CdS at room temperature, 3.88×10^{-19} J, (14)
- r = particle radius in meters,
- m_e^* = effective mass of conduction band electron in CdS, 1.73×10^{-31} kg, (15)
- m_h^* = effective mass of valence band hole in CdS, 7.29×10^{-31} kg, (15)
- ε = relative permittivity of CdS, 5.7, (15)
- h is Planck's constant and ε_o is the permittivity of a vacuum

The nanoparticle band gap energy (E_g^{nano}) can be approximated from the cutoff wavelength in the absorbance spectrum. The cutoff wavelength is the x-intercept of a line drawn through the absorbance data points located near the onset of absorbance. Equation 2 is then rearranged to solve for the particle radius. A spectrum typical of 5.4 nm diameter CdS particles prepared according to this laboratory experiment procedure is shown in Figure 1.

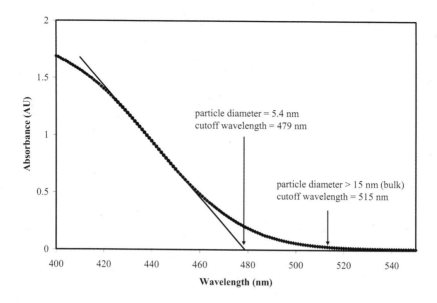

Figure 1. UV/visible spectrum of CdS nanoparticles showing cutoff wavelengths and corresponding diameters for nanosized and bulk CdS particles.

In this experiment, micelles form in a water-in-oil emulsion. The surfactant has a long hydrocarbon tail that extends into the hydrophobic solvent and a cationic head group that is oriented towards the aqueous phase. Each micelle contains nanometer-sized droplets of either aqueous $Cd^{2+}_{(aq)}$ or $S^{2-}_{(aq)}$ ions. The surfactant "shell" is not rigid so when two micelles collide, their contents mix as the surfactant molecules rearrange themselves to form two new micelles. In this manner, micelles serve as "nanoreactors". Mixing Cd^{2+} and S^{2-} ions leads to the precipitation of solid CdS particles with their size limited by the micelle dimensions. The arrangement of the surfactant around the aqueous phase reduces contact between nanoparticles by creating a steric barrier and through electrostatic repulsion between micelles. CdS nanoparticle emulsions prepared by this method are stable for at least several hours. A brief but adequate discussion of surfactants is included in many general chemistry textbooks. Students should have a good understanding of intermolecular forces so that they can understand why surfactant molecules aggregate to form micelles.

Procedure

Prior to the laboratory session, the instructor should prepare aqueous stock solutions of $CdCl_2$ and Na_2S. The sodium sulfide solution should be fresh but the cadmium chloride solution can be prepared well in advance. Students prepare two batches of solvent by mixing hexanes with 1-pentanol. Next, they add cetyltrimethylammonium bromide (CTAB) surfactant to each. CTAB is only slightly soluble in the organic solvent but upon the addition of a few drops of aqueous salt solution, CTAB dissolves completely. This creates a transparent water-in-oil emulsion in which the aqueous phase is surrounded by CTAB molecules and dispersed within the pentanol-hexanes solution. Students mix the two emulsions to create a translucent yellow solution. CdS nanoparticles are typically 5–6 nm in diameter and show a shift of approximately 40 nm in their absorbance spectrum compared to bulk cadmium sulfide. A slight variation in particle size was observed depending on the relative amounts of hexanes, 1-pentanol and water used. Students prepare bulk CdS by simply mixing equal portions of the aqueous cadmium chloride and sodium sulfide solutions. Careful observers will note that bulk CdS particles initially appear yellow before changing to their characteristic orange color and precipitating. A comparison between the nanoparticle CdS emulsion and the bulk aqueous suspension shows that the two materials are clearly different.

Students should perform this procedure in a fume hood due to the organic solvent vapors. Proper disposal of all chemicals is important since cadmium compounds are toxic and carcinogenic.

Applications

Many current and future applications of CdS nanoparticles are based on their high surface area and absorption of visible light to produce charge carriers (e^-_{cb} and h^+_{vb}). For a given mass, smaller particles have a much greater total surface area than bulk particles so nanoparticles can provide many more surface adsorption sites for catalytic reactions. Cadmium sulfide has been extensively studied because its band gap allows the material to absorb visible light. This is an important characteristic for solar energy driven processes because the solar spectrum at sea level contains much more visible than ultraviolet light. Photogenerated e^-_{cb} and h^+_{vb} can cause reduction and oxidation reactions, respectively, with adsorbed molecules. Since these reactions occur in combination, CdS is a photocatalyst.

These properties lead to a variety of applications. Inclusion of CdS nanoparticles can make photovoltaic devices, which convert solar energy into electricity, more efficient (*16*). They can be effective photocatalysts for the degradation of many organic pollutants (*17-19*). Composite materials that incorporate CdS nanoparticles can generate H_2 gas from water when illuminated with visible light (*20, 21*). Conduction band electrons in cadmium sulfide nanoparticles are capable of chemically reducing CO_2 to form small organic molecules which could be used as a source of alternative fuel (*22*).

Particle size not only affects the color of the semiconductor but also its fluorescence spectrum. This is observed in CdS and to a greater extent in CdSe nanoparticles. Such compounds are useful as stains for biological cells. Different nanoparticles can be directed into various regions of a cell by functionalizing the particle's surface. When irradiated with UV light, the nanoparticles make cell features easier to see under a fluorescence microscope. This is currently done using organic dyes which degrade quickly inside the cells. Nanoparticles have a longer useful lifetime and provide researchers with more time to study the cell. Certain functional groups on the nanoparticle surface can bind to a virus or tumor cell in the body. The functionalized nanoparticles will accumulate around the target, and their fluorescence makes the malignant cells easier to locate (23).

Magnetite Nanoparticles (Ferrofluid)

Ferrofluids are a visually impressive demonstrations of nanotechnology. Students in Florida Tech's freshman nanotechnology laboratory course consistently rate this experiment as their favorite. It requires no laboratory instrumentation, just glassware and a strong magnet. Students create a suspension containing mixed iron oxide particles coated with charged surfactant molecules to prevent agglomeration. After separating the ferrofluid from the solvent, they explore how the fluid behaves in a magnetic field (24, 25).

This experiment provides the opportunity to discuss magnetism, a topic not often addressed in the general chemistry laboratory. Surfactants play an important role in this reaction so students should understand concepts such as electrostatic charge repulsion and intermolecular forces. Instructors might illustrate the ferrofluid's behavior in a magnetic field by showing the analogous and more familiar effect of a magnet on iron filings in a dish.

A simple but important topic related to ferrofluid synthesis is stoichiometry. The 2:1 molar ratio of Fe(III) to Fe(II) shown in reaction 2 below must be strictly observed. Deviations from this ratio result in a mixture of magnetite and other iron oxides that do not respond to an applied magnetic field.

Background

A ferrofluid is a liquid that responds to a magnetic field. This response is due to the presence of ferrimagnetic nanoparticles suspended in a solvent. Ferrimagnetic materials contain regions within the lattice, called domains, in which one spin state is more common than the other. This leads to a strong interaction with the applied magnetic field. In the magnetite unit cell, the spins of the unpaired electrons from the two iron(III) ions are antiparallel to each other, meaning that they cause no net interaction with the applied magnetic field. Within a domain, the spins of unpaired electrons of the iron(II) ions are aligned parallel with each other, creating a net magnetization in the presence of an applied magnetic field. This magnetization is short lived since the size of the domains is on the order of the particles themselves.

Magnetite (Fe_3O_4) forms according to reaction 3:

$$2\ FeCl_{3\ (aq)} + FeCl_{2\ (aq)} + 8\ NH_{3\ (aq)} + 4\ H_2O_{\ (l)} \rightarrow Fe_3O_{4\ (s)} + 8\ NH_4Cl_{\ (aq)} \quad (3)$$

Bulk Fe_3O_4 crystals form slowly, allowing students time to add an ionic surfactant that limits the particle size to approximately 15 nm in diameter. The anions adsorb to the magnetite surface after the Fe_3O_4 particles have formed. Electrostatic repulsion reduces the number of collisions between the particles.

Water molecules in the solution are also attracted to the surfactant molecules that surround the magnetite particles. This attraction prevents the Fe_3O_4 particles from separating from the solution when a magnet is brought near. Instead, the magnetite particles "drag" the solvent towards the magnet. Since the solution moves with the ferrimagnetic particles, it appears that the whole liquid is magnetic. A strong magnetic field also increases the viscosity of the ferrofluid, holding the liquid in place. As a result, ferrofluids will "spike" when a strong magnet is held near because the fluid follows the magnetic field lines. This behavior is illustrated in Figure 2. It should be pointed out that heating a ferrimagnetic solid above its melting point would not create a magnetic fluid because the heat will provide enough energy to randomize the orientation of the electron spins before the solid melted (24).

Figure 2. Pictures of a ferrofluid in the presence (top) and absence (bottom) of a magnetic field provided by the magnet underneath the dish.

As is the case in the synthesis of CdS nanoparticles, a surfactant (tetramethylammonium hydroxide) is used to prevent magnetite nanoparticle agglomeration. The hydroxide anions adsorb to the magnetite surface after the Fe_3O_4 particles have formed and the cations form a loose layer around them. Electrostatic repulsion reduces the number of collisions between the particles.

Procedure

The instructor can prepare stock solutions of HCl and NH_3 prior to the laboratory session. Students add iron(II) and iron(III) chloride in separate HCl solutions and then mix them together. Magnetite precipitates when ammonia is added dropwise to the iron chloride solution. Students use a magnet to collect the solid and separate it from the solution. They add tetramethylammonium hydroxide to delay particle growth. This creates a thick, black, ferrofluid liquid containing 10 – 20 nm diameter Fe_3O_4 particles that are stable for at least several hours. After completing the synthesis, students observe the behavior of the magnetite colloid in the presence of a magnetic field. Instructors should consult the original procedure for ideas of what students can do with their ferrofluid (24, 25). It is advised that students do not allow the magnet to come into direct contact with the ferrofluid since the solution is difficult to remove. Over time, the water will evaporate, leaving behind solid magnetite. If a magnet is placed under a dish of ferrofluid overnight, the magnetite in the "spikes" will solidify, making an interesting momento from the experiment.

Ammonia fumes necessitate ferrofluid synthesis in a fume hood. Ferrofluids stain skin and clothing so lab coats and gloves are recommended.

Applications

Ferrofluids are currently used in a wide variety of fields, from aerospace to sculpture to anti-counterfeiting. Most applications require ferrofluids with greater stability than those prepared here. Commercially available ferrofluids contain particles suspended in a nonvolatile organic solvent instead of water. The manufacturing parameters are customized for different applications. The original and still most common use of ferrofluids is as a seal between moving machine parts, such as a rotating shaft (26). A ring of permanent magnets creates a magnetic field around the shaft and the ferrofluid is held in place by the magnets, forming a seal. Ferrofluids have significant advantages over other seal materials, including long operational lifetime (over 10 years), consistent performance over a range of applied pressures, ease of manufacture, and low permeability of gases. These qualities make ferrofluid-based seals appropriate for use in satellite gyroscopes, automobiles, oil refineries, computer hard drives, and various manufacturing processes in the chemical, textile, and biotechnology industries. College students might be more interested in ferrofluids' use as a magnetically driven damper and a heat sink in their stereo speakers. The viscosity of a ferrofluid, and therefore its damping ability, can be controlled by a magnetic field. Ferrofluids are a better conductor of heat than air (27).

By placing a drop of their ferrofluid on a small piece of paper, students can observe another application of ferrofluids – as a magnetic ink. Ferrofluids stain the paper and the magnetite nanoparticles become embedded in the cellulose paper material. The attraction of the magnetite towards a magnet can lift the paper. Magnetic inks such as this are part of the anti-counterfeiting measures found in U. S. currency (24).

Probably the most impressive use of ferrofluids is as a sculpting material. By combining artistic creativity and knowledge of magnetism, artists can create amazing structures that change as the magnetic field varies. Sachiko Kodama's "Morpho Towers – Two Standing Spirals" adds synchronized music to the ever changing sculpture (*28*). In another case, complex designs reminiscent of Native American artwork were created unexpectedly by researchers at MIT's electrical engineering and computer science department (*29, 30*).

Ferrofluids may be useful in the medical field. Pharmaceutical drug molecules can be bound to the surface of magnetic nanoparticles and guided to a specific location in the body using an external magnetic field (*31*). Using the same method, they could also treat cancer by coating the particles with a substance that is toxic to tumor cells (*32*). Another way to kill a malignant cell is through ferrofluid hyperthermia. Coating the nanoparticles with a biocompatible substance allows them to enter a cell and, once inside, an oscillating magnetic field causes the ferrofluid nanoparticles to vibrate. This motion generates enough heat to destroy the cell (*33*).

Gold Nanoparticles and Self-Assembled Monolayers

Students in Florida Tech's Nanoscience and Technology laboratory synthesize gold nanoparticles as a guided inquiry experiment. The gold colloid solution changes color depending on the type of solute (molecular or ionic) added. Students are presented with a variety of solutes from which to choose and at the end of the laboratory session, they pool their data. Based on the class's results, they formulate a hypothesis for why certain solutes cause a change in color and how the nanoparticles are affected by the change in the solution environment. The color change is easily visible and does not require a spectrophotometer. General chemistry topics that are involved in this experiment include redox reactions, light absorption, reaction rates, and electrolytes.

Two methods are available for the synthesis of gold nanoparticles from gold(III) ions, as $HAuCl_4$. In one procedure, citrate acts as a reducing agent to form 13 nm diameter gold nanoparticles (*34, 35*). This reaction has been successfully performed in many general chemistry laboratory courses and Florida Tech's NST lab. A recently published article in the *Journal of the American Chemical Society* provides an alternative photochemical route in which gold(III) is reduced to metallic gold by an organic radical that forms upon exposure to UV light from the sun (*36, 37*).

In another experiment, students prepare layers of gold nanoparticles on glass (*38*). With each additional layer, the color of the gold film changes from pink (first layer) to the characteristic gold color (fifteenth layer). This experiment was originally designed for a physical chemistry laboratory course but it appears to be suitable for the general chemistry laboratory after making minor modifications. It could also be an effective demonstration to show a class how a material's properties progressively change from the nanoscale to the bulk.

Students can prepare a gold colloid solution on the bench top. Layering gold nanoparticles is best performed in fume hood while wearing gloves due to the harmful chemicals used to link the gold layers together.

Background

Assuming that the citrate ions are completely oxidized by tetrachloroaurate(III), reaction 4 shows the reduction-oxidation reaction that causes the formation of the red gold colloid (Cl⁻ omitted):

$$12 \text{ Au}^{3+}_{(aq)} + 2 \text{ C}_6\text{H}_5\text{O}_7^{3-}_{(aq)} + 10 \text{ H}_2\text{O}_{(l)} \rightarrow 12 \text{ Au}_{(s)} + 12 \text{ CO}_{2(g)} + 30 \text{ H}^+_{(aq)} \quad (4)$$

Citrate ($C_6H_5O_7^{3-}$) not only reduces the gold cations but adsorbs to the gold surface. The buildup of negative charge on the surface creates an electrostatic repulsive force between particles. This keeps the particles from growing by slowing the rate that particles collide. Similar methods are employed by surfactants in the laboratory experiments described earlier. However, a notable difference is that the citrate ions do not provide a significant steric barrier, as demonstrated by the addition of ions to the gold colloid solution.

Addition of non-reactive ions to a solution increases its dielectric constant. As a result, the electrostatic attractive and repulsive forces between ions are both diminished (recall that the dielectric constant appears in the denominator of Coulomb's law). Two species with negative charges, such as citrate-complexed gold nanoparticles, will not strongly repel each other in a high ionic strength solution. Less repulsion allows the particles to approach each other more closely, increasing the likelihood of a collision. The rate of collisions, and therefore the rate of agglomeration, increases when electrolytes are present in the gold colloid solution. This is known as the primary kinetic salt effect. A full mathematical treatment can be found in most physical chemistry textbooks. A qualitative analogy that students seem to easily understand is the attraction (or repulsion) between two people. It would be easy for the pair to approach (or stay away from) each other in an empty room but more difficult to find each other (or bump into each other) in a crowded room.

The gold colloid appears red due to its surface plasmon resonance band. This absorption peak, centered at approximately 520 nm, shows the energy of light that is absorbed by electrons on the gold particle surface. Its position is affected by the size of the gold particles. As the particles become larger, the intensity of the 520 nm plasmon resonance band peak decreases but never completely disappears while a broad absorption is observed at approximately 750 nm. The absorption of 750 nm becomes more intense as more gold particles grow (due to more electrolytes in the solution) until it is more intense than the 520 nm band. This results in a solution that appears blue-gray colored. The two different spectra can be seen in Figure 3 for a gold colloid solution with and without NaCl added.

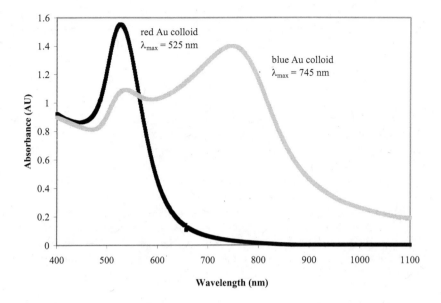

Figure 3. UV/visible spectra of red (no NaCl added) and blue (with NaCl) gold colloid solutions with their plasmon resonance band's wavelength of maximum absorbance indicated.

Similar color changes are observed as multiple layers of gold nanoparticles are deposited on a glass slide. The observed colors are pink (one layer), blue (two layers), gray (five layers), and eventually the gold color typical of bulk gold (fifteen layers). Since each layer of gold nanoparticles is separated by a layer of crosslinking molecules, the gold nanoparticles do not have to be directly bonded together in order to show a change in the surface plasmon resonance.

Surface plasmon resonance is due to the absorption of light at the surface of a metal, typically gold or silver. The effect extends only up to a few hundred nanometers into the material. At the surface, the electrons around the metal nuclei behave as a plasma and can be excited by light that couples with the frequency of their collective oscillation (resonance). Due to the confinement of the electrons by the physical dimensions of the nanoparticle, the excitation energy increases as the particle size decreases. A similar effect was described earlier for excitons in semiconductor nanoparticles.

Procedure

Prior to the laboratory session, the instructor should prepare a stock $HAuCl_4$ solution and various electrolytic and nonelectrolytic solutions for analysis. The tetrachloroaurate(III) solution is stable for at least a year. The number of analysis solutions should exceed the number of lab groups performing the experiment so that each group collects different data. Strong and weak acids, bases and non-ionic compounds such as table sugar are good examples of solutes. The effect of these solutions on the gold nanoparticles should be tested beforehand to make sure that a color change occurs as expected. Occasionally, it has been found that a solute causes an undesirable side reaction that interferes with determining the gold colloid solution's color. These solutions should be labeled with the solute's name and chemical formula.

To prepare gold nanoparticles via the citrate method, each laboratory group dilutes a portion of the gold(III) stock solution and brings it to a boil, periodically replacing water that has vaporized. Students add sodium citrate to reduce the gold(III) ions to metallic gold nanoparticles. Upon reduction, the solution color becomes a deep red. After the solutions cool, students pour 2 mL of the gold colloid solution into small vials or plastic cuvettes. Students then add several drops of different analysis solutions and record their observations.

At the end of the laboratory session, students can share their data. Results should show that strong electrolytes cause the red Au colloid to turn blue and weak or nonelectrolytes have no effect on the colloid's color. The instructor can then ask students to hypothesize the reason for this behavior.

Formation of a gold self-assembled monolayer on glass is accomplished by first treating the glass slide with 3-aminopropyltriethoxysilane. Due to the hazards of working with silanes, the instructor should treat the glass slides before the laboratory session. When students dip the slides in a gold colloid solution, the amine functional groups on the slide bind the gold nanoparticles, creating a monolayer of gold on glass. The slide can then be treated with a 2-mercaptoethyamine solution which acts as a crosslinking agent. Immersing the slide again in the gold colloid solution deposits another layer of gold on top of the amine-functionalized gold monolayer. Students can repeat these steps to form many more layers.

Applications

Gold nanoparticles may be the first synthetic nanoparticles created by man. They have been used for centuries in ceramic pottery glazes and tinting for churches' stain glass windows. Depending on the size and environment of the gold nanoparticles, they can have a variety of colors (*39*).

A more recently developed application uses the easily observed red-blue color change of the gold colloid to detect a specific sequence of DNA (*23*). Gold nanoparticles are prepared using the same procedure described above and separated into two batches. Portions of a certain DNA sequence of interest (probe DNA) are selected and bound to one batch of gold nanoparticles. Portions of another, different probe DNA sequence are attached to nanoparticles in the other batch. The particular DNA sequences are based on their ability to bond to the ends of a target DNA strand of interest that may be in the solution. Gold nanoparticles will become closely spaced if both of their pieces of probe DNA attach to the ends of the target DNA. The closely spaced gold particles will turn the solution's color from red to blue. No color change will occur if only one probe or neither probe DNA binds to the target DNA.

Silver Nanoparticles

A variety of simple methods exist for synthesizing silver nanoparticles. Silver(I) cations are reduced by borohydride (*40, 41*), sugars (*42*) or a plant extract (*43*). Reduction of silver by any of the above methods can be performed by general chemistry students with the sugar reductant method being the simplest. An advantage of synthesizing silver nanoparticles compared to gold nanoparticles is price – silver nitrate is significantly less expensive than hydrogen tetrachloroaurate(III).

The use of sugars as a reductant is particularly interesting because the same general method, known as Tollens' reaction, is used in the popular demonstration of the deposition a silver mirror (*44*), a demonstration of contact angles on a silver surface (*45*) and of course to test for the presence of aldehydes in organic chemistry. This procedure allows instructors to show how slight modifications in the experimental procedure can lead to different forms of metallic silver. The silver mirror demonstration can be a useful demonstration to show how the color of silver on the glass surface changes as layers of silver atoms are deposited. The glass initially turns a yellow-brown hue, similar to the color of Ag colloids formed by these procedures.

Background

Tollens' reaction is used to detect the presence of aldehyde groups. Silver(I), in the form of the silver diamine cation, is reduced to silver metal while the aldehyde is oxidized to a carboxylic acid. The metallic silver deposits on the inner surface of the glass container. Oxidation and reduction reactions that occur during Tollens' reaction are shown in reactions 5 and 6:

$$RCHO_{(aq)} + 2\ OH^-_{(aq)} \rightarrow RCOOH_{(aq)} + H_2O_{(l)} + 2\ e^- \qquad (5)$$

$$2\ [Ag(NH_3)_2]^+_{(aq)} + 2\ e^- \rightarrow 2\ Ag_{(s)} + 4\ NH_{3\ (aq)} \qquad (6)$$

Under these conditions, ketones are not oxidized. This is an important distinction since some sugars, called reducing sugars, each contain an aldehyde functional group, but the nonreducing sugars do not. The saccharides used in this experiment are all reducing sugars. Table sugar, sucrose, is the most notable nonreducing sugar.

In this experiment, silver(I) is reduced to silver metal but the particles remain small ($\leq$ 50 nm diameter). Silver particle size is controlled by the solution pH, type of saccharide used and concentration of ammonia. Each of these parameters affects the redox potentials of the two half reactions shown in Equations 5 and 6, which in turn influence the reaction rate. Under the proper experimental conditions, many small silver metal clusters form rapidly. This is followed by slower growth of the clusters into nanoparticles. Silver metal atoms formed subsequently are distributed among the many growing particles, creating many very small particles (*42*). Under other conditions, such as those used to perform Tollens' test for aldehydes, a few clusters form initially and then grow to create bulk silver.

Small silver nanoparticles are stabilized by ionic species that adsorb to their surface, much like the citrate ions prevent rapid agglomeration of gold nanoparticles. The ionic repulsion experienced by the silver particles are likely due to carboxylic acids that form after the oxidation of aldehydes. Upon the addition of other ionic species, the silver nanoparticles approach each other more easily and grow in size. Students observe a change in color from yellow to green as the plasmon resonance band shifts to longer wavelengths.

Procedure

The instructor should prepare all solutions before the laboratory session. Sugar solutions need to be made no more than one day in advance because bacteria will grow within the solution. Students add a silver nitrate solution to a test tube, followed by the addition of a few drops of sodium hydroxide. A brownish silver hydroxide precipitate immediately forms. This dissolves by adding a few drops of ammonia solution. Silver nanoparticles slowly form after any of a variety of sugars (glucose, maltose, galactose, lactose) are added (*42*).

Different sugars yield particles of different average sizes and therefore slightly different solution colors. Using maltose as the reducing agent generates the smallest silver nanoparticles (25 nm diameter) and a yellow solution with a plasmon resonance band centered at approximately 415 nm. Larger particles (50 nm diameter) make the solution appear brownish-yellow colored ($\lambda_{max} \cong 450$ nm) when glucose is the reducing agent. Solutions containing larger average particles also show a greater range of sizes based on the half width of the visible spectrum absorbance peak. The borohydride procedure is recommended to form nanoparticles with diameters less than 20 nm ($\lambda_{max} \cong 400$ nm) (*40, 41*). The plant extract method yields particles with an average diameter of 30 nm (*43*).

Students can observe particle aggregation due to the kinetic salt effect upon adding a strong electrolyte, similar to the gold nanoparticle experiment. Chlroide salts are not recommended since AgCl can precipitate, interfering with observations. The absorbance at 400 – 450 nm decreases and a broad absorption band appears at ~ 650 nm. The salted solution has a blue-green color. Without the addition of salt, the silver colloid is stable for at least several days. Interestingly, the nanoparticles formed by using borohydride instead of sugars turn brown when salt is added (*40, 41*). Use of the germanium extract as a reducing agent reportedly produces silver nanoparticles in a red-brown colored solution and a broad plasmon resonance peak. Adding a salt changes the colloid's color to yellow (*43*).

Applications

People have known about the infection-fighting ability of silver salts for many centuries and doctors today use ionic silver to prevent bacteria growth. Silver compounds have been incorporated in plastics, textiles and on the surfaces of many items used in medical procedures (*46*). Although the mechanism is not fully understood, evidence suggests that silver cations react with sulfur-containing amino acids and DNA (*47*). Silver nanoparticles can pass through the bacterium cell wall and show biocidal activity against both gram positive and gram negative bacteria. The effectiveness of a silver colloid appears to be inversely related to the size of the Ag particles. This could be simply due to the greater surface area of smaller particles or their greater number, since both factors would lead to more interactions with bacteria cells. Alternatively, smaller particles could enter the cells more easily. Atoms of elemental silver might be oxidized to Ag^+ within or around the cell, meaning that silver cations are the actual anti-bacterial agent even when metallic silver is used. Silver nanoparticles within the cell do not appear to trigger the cell's defense mechanism when it detects toxins, as silver cations do. A full explanation of the anti-bacterial properties of silver nanoparticles is not clear (*42, 48*). Given the increasing threat of antibiotic-resistant bacteria, scientists are interested in learning more about the biological effects of silver nanoparticles.

Medieval glassworkers made yellow stained glass for church windows by adding silver compounds to the glass before heating. The silver cations were reduced to metallic silver in a redox reaction involving Fe^{3+} or other ions found in the aluminum-silicate glass (*49*). Students can prepare yellow stained glass-type material (*41*).

Nickel Nanowires

Nickel nanowires are prepared using a nanoscale template to control their dimensions (*50-52*). Prior to performing this experiment, students should learn about electrochemistry and the components of an electrochemical cell in their general chemistry lecture class so that they understand the procedure. Freshmen in Florida Tech's Nanoscience and Technology laboratory course found this experiment to be the most challenging experiment due to the number of steps involved. Some students did not achieve satisfactory results due to their inattention to details.

Background

Template synthesis has been a common route for preparing nanomaterials for many years. Similar to the micelles that act as nanoreactors for CdS nanoparticle synthesis, a template restricts the growth of the nanomaterial in one or more dimensions. Unlike micelles, the template used in this experiment is pre-fabricated. Templates can be synthesized or naturally occurring. Aluminum oxide filters and zeolites, respectively, are examples of these types of templates. The possible dimensions of the resulting nanomaterial depend only on the availability of templates with different sized and shaped pores. This can lead to a greater variety of nanoscale products. For that reason, nanoscale templating will be a likely method for the production of parts for nanomachines developed in the future. A review of different carbon-based nanostructures provides an overview of some current templating techniques (*53*).

Procedure

Students construct an electrochemical cell with a nickel wire cathode and a copper plate as the anode. One side of an aluminum oxide filter with 20 nm diameter channels is exposed to a nickel plating solution when attached to the copper electrode. Gallium-indium eutectic paint is applied to the other side of the filter in order to maintain electrical contact with the electrode. Students connect the electrodes to a 1.5 V battery for approximately 30 minutes as the nickel cations are reduced to nickel metal within the filter pores. Nickel nanowires grow at a rate of approximately 1 micrometer per minute.

Nanowires are separated from the disk by first dissolving the tape adhesive in acetone and then dissolving the eutectic paint with concentrated nitric acid. The aluminum oxide disk is fragile and students should exercise great care when handling it. Students dissolve the aluminum oxide filter in a sodium hydroxide solution, releasing the nanowires from the template. A magnet attracts the nanowires to the bottom of the beaker while students decant the NaOH solution. They should wash the nanowires several times with deionized water.

Nickel nanowires can be stored indefinitely under water in a vial. Nickel is a ferromagnetic material and so the wires can be manipulated with a magnet. Like the magnetite ferrofluid, the nanowires will position themselves along the magnetic field lines. However, they are not suspended in water and so the liquid medium itself is not affected by the magnet. Students can observe the nanowires using a scanning electron microscope or even an optical microscope since the wires aggregate into larger, rod-shaped clusters.

Students should wear goggles, lab coats, and gloves and use extra caution when handling the concentrated acid and base solutions. Steps involving those solutions are performed in a fume hood.

Applications

Nanowires can be used as chemical sensors. A nanowire grown between two electrodes can detect the presence of nanogram quantities of solutes (*54*). When the solute adsorbs to the nanowire, the conductance between the electrodes changes by an amount that is related to the substrate's adsorption strength. The absorbed molecules or ions can be identified by changing the applied voltage and monitoring the change in conductance.

Data is recorded to a computer hard drive by applying a magnetic field that manipulates particles within the disk medium. Smaller particles allow for a greater density of information to be stored. Nickel nanowires respond to an applied magnetic field, making them an attractive recording material for future data storage devices (*55*).

Resources

More educational resources, many freely available on the Internet, will become available as nanotechnology education matures. Chemical education journals such as *The Journal of Chemical Education* and *The Chemical Educator* feature frequent nanotechnology-themed articles describing lecture topics, demonstrations and laboratory experiments. Since nanotechnology is an interdisciplinary field, education journals for biology, engineering, physics and materials science can provide articles of interest. The new *Journal of Nano Education* features relevant information from many fields of science and engineering. The University of Wisconsin Madison's MRSEC education website provides a video lab manual for many of the experiments presented in this chapter and additional background information concerning nanotechnology and materials science. A recently published book, Nanoscale Science and Engineering Education, provides extensive coverage of this field and includes an in-depth description of Florida Tech's NST laboratory course (*56*). The National Science Digital Library is a searchable database with links to many nanotechnology-themed web sites.

New nanotechnology experiments for freshmen can be found in recent issues of the science and engineering research literature. What were cutting edge syntheses a few years ago can now be performed by freshmen students in their general chemistry lab course, as is the case for the synthesis of CdS nanoparticles described earlier in this chapter. Many books have been published that introduce students and the general public to all areas of nanotechnology, from science to its impact on business and society.

References

1. Roco, M. C. *Int. J. Eng. Ed.* **2002**, *18*, 488.
2. Shapter, J. G.; Ford, M. J.; Maddox, L. M.; Waclawik, E. R. *Int. J. Eng. Ed.* **2002**, *18*, 512.
3. Roco, M. C. *Nat. Biotechnol.* **2003**, *21*, 1247.
4. Condren, S. M.; Breitzer, J. G.; Payne, A. C.; Ellis, A. B.; Windstrand, C. G.; Kuech, T. F.; Lisensky, G. C. *Int. J. Eng. Ed.* **2002**, *18*, 550.
5. Roco, M. C. *J. Nanopart. Res.* **2003**, *5*, 181.
6. Porter, L. A., Jr. *J. Chem. Educ.* **2007**, *84*, 259.
7. Colvin, V. L. *Nat. Biotechnol.* **2003**, *21*, 1166.
8. Morrissey, S. *Chem. Eng. News* **2005**, *83*, 46.
9. Curri, M. L.; Agostiano, A.; Manna, L.; Monica, M. D.; Catalano, M.; Chiavarone, L.; Spagnolo, V.; Lugara, M. *J. Phys. Chem. B* **2000**, *104*, 8391.
10. Winkelmann, K.; Noviello, T.; Brooks, S. *J. Chem. Educ.* **2007**, *84*, 709.
11. Hansen, P.; Lisensky, G. Synthesis of Cadmium Sulfide Nanoparticles. http://www.mrsec.wisc.edu/Edetc/nanolab/CdS (June, 2008).
12. Kippeny, T.; Swafford, L. A.; Rosenthal, S. J. *J. Chem. Educ.* **2002**, *79*, 1094.
13. Brus, L. *J. Phys. Chem.* **1986**, *90*, 2555.

14. *CRC Handbook of Chemistry and Physics*, 73rd ed.; Lide, D. R., Ed. CRC Press: Boca Raton, FL, 1992; pp 12-94 to 12-98.
15. Brus, L. E. *J. Chem. Phys.* **1984**, *80*, 4403.
16. Yang, S.; Wen, X.; Zhang, W.; Yang, S. *J. Electrochem. Soc.* **2005**, *152*, G220.
17. Yin, H.; Wada, Y.; Kitamura, T.; Yanagida, S. *Environ. Sci. Technol.* **2001**, *35*, 227.
18. Warrier, M.; Lo, M. K. F.; Monbouquette, H.; Garcia-Garibay, M. A. *Photochem. Photobiol. Sci.* **2004**, *3*, 859.
19. Han, M. Y.; Huang, W.; Quek, C. H.; Gan, L. M.; Chew, C. H.; Xu, G. Q.; Ng, S. C. *J. Mater. Res.* **1999**, *14*, 2092.
20. Stroyuk, A. L.; Korzhak, A. V.; Raevskaya, A. E.; Kuchmii, S. Y. *Theor. Exp. Chem.* **2004**, *40*, 1.
21. Hirai, T.; Bando, Y. *J. Colloid Interface Sci.* **2005**, *288*, 513.
22. Kamat, P. V.; Murakoshi, K.; Wada, Y.; Yanagida, S. In *Nanostructured Materials and Nanotechnology*, Nalwa, H. S., Ed. Academic Press: San Diego, CA, 2002; pp 130-182.
23. Alivisatos, A. P. *Sci. Amer.* **2001**, *285*, 67.
24. Berger, P.; Adelman, N. B.; Beckman, K. J.; Campbell, D. J.; Ellis, A. B.; Lisensky, G. C. *J. Chem. Educ.* **1999**, *76*, 943.
25. Breitzer, J.; Lisensky, G. Synthesis of Aqueous Ferrofluid. http://www.mrsec.wisc.edu/Edetc/nanolab/ffexp/index.html (June, 2008).
26. Ochonski, W. *Machine Design* **2005**, *77*, 118.
27. Tatarunis, S.; Klasco, M. *Voice Coil* **2005**, p 1.
28. Kodama, S. Morpho Towers - Two Rotating Spirals. http://www.kodama.hc.uec.ac.jp/spiral/ (July, 2007).
29. Rhodes, S.; Perez, J.; Elborai, S.; Lee, S.-H.; Zahn, M. *J. Magn. Magn. Mater.* **2005**, *289*, 353.
30. Lorenz, C.; Zahn, M. *Phys. Fluids* **2003**, *15*, S4.
31. Alexiou, C.; Jurgons, R.; Schmid, R.; Hilpert, A.; Bergemann, C.; Parak, F.; Iro, H. *J. Magn. Magn. Mater.* **2005**, *293*, 389.
32. Hilger, I.; Fruhauf, S.; Linss, W.; Hiergeist, R.; Andra, W.; Hergt, R.; Kaiser, W. A. *J. Magn. Magn. Mater.* **2003**, *261*, 7.
33. Giri, J.; Ray, A.; Dasgupta, S.; Datta, D.; Bahadur, D. *Bio-Med. Mater. Eng.* **2003**, *13*,, 387.
34. McFarland, A. D.; Haynes, C. L.; Mirkin, C. A.; Van Duyne, R. P.; Godwin, H. A. *J. Chem. Educ.* **2004**, *81*, 544A.
35. Lisensky, G. Citrate synthesis of gold nanoparticles. http://www.mrsec.wisc.edu/Edetc/nanolab/gold/index.html (July, 2007).
36. McGilvray, K. L.; Decan, M. R.; Wang, D.; Scaiano, J. C. *J. Am. Chem. Soc.* **2006**, *128*, 15980.
37. Lisensky, G. Photochemical synthesis of gold nanoparticles. http://www.mrsec.wisc.edu/Edetc/nanolab/gold/index2.html (July, 2007).
38. Oliver-Hoyo, M.; Gerber, R. W. *J. Chem. Educ.* **2007**, *84*, 1174.
39. Ratner, M.; Ratner, D., *Nanotechnology: A Gentle Introduction to the Next Big Idea*. Prentice Hall: Saddle River, NJ, 2003; pp 12-14.
40. Solomon, S. D.; Bahadory, M.; Jeyarajasingam, A. V.; Rutkowsky, S. A.; Boritz, C.; Mulfinger, L. *J. Chem. Educ.* **2007**, *84*, 322.

41. Ng, S.; Johnson, C. Synthesis of Colloidal Silver. http://www.mrsec.wisc.edu/Edetc/nanolab/silver/index.html (July, 2007).
42. Panacek, A.; Kvitek, L.; Prucek, R.; Kolar, M.; Vecerova, R.; Pizurova, N.; Sharma, V. K.; Nevecna, T. j.; Zboril, R., *J. Phys. Chem. B* **2006**, *110*, 16248.
43. Richardson, A.; Janiec, A.; Chan, B. C.; Crouch, R. D. *Chem. Educator* **2006**, *11*, 331.
44. Shakhashiri, B. Z. In *Chemical Demonstrations: A Handbook for Teachers of Chemistry*, University of Wisconsin Press: Madison, WI, 1992; Vol. 4, pp 240-243.
45. Lisensky, G. Octadecanethiol Monolayer on Silver. http://www.mrsec.wisc.edu/Edetc/nanolab/Agthiol/index.html (July, 2007).
46. Weber, D. J.; Ruala, W. A. In *Disinfection, Sterilization, and Preservation*, 5th ed.; Block, S. S., Ed. Lippincott Williams and Wilkins: Philadelphia, PA, 2001; pp 418-423.
47. Liau, S. Y.; Read, D. C.; Pugh, W. J.; Furr, J. R.; Russell, A. D. *Lett. Appl. Microbiol.* **1997**, *25*,, 279.
48. Morones, J. R.; Elechiguerra, J. L.; Camacho, A.; Holt, K.; Kouri, J. B.; Ramirez, J. T.; Yacaman, M. J. *Nanotechnol.* **2005**, *16*, 2346.
49. Jembrih-Simbuerger, D.; Neelmeijer, C.; Schalm, O.; Fredrickx, P.; Schreiner, M.; De Vis, K.; Maeder, M.; Schryvers, D.; Caen, J. *J. Anal. At. Spectrom.* **2002**, *17*, 321.
50. Bentley, A. K.; Farhoud, M.; Ellis, A. B.; Lisensky, G. C.; Nickel, A.-M. L.; Crone, W. C. *J. Chem. Educ.* **2005**, *82*, 765.
51. Lisensky, G. Synthesis of nickel nanowires: beaker. http://www.mrsec.wisc.edu/Edetc/nanolab/nickel/index.html (July, 2007).
52. Lisensky, G. Synthesis of nickel nanowires: syringe. http://www.mrsec.wisc.edu/Edetc/nanolab/nickel/index2.html (July, 2007).
53. Kyotani, T. *Bull. Chem. Soc. Jpn.* **2006**, *79*, 1322.
54. Tao, N., *J. Chem. Educ.* **2005**, *82*, 720.
55. Pignard, S.; Goglio, G.; Radulescu, A.; Piraux, L.; Dubois, S.; Declemy, A.; Duvail, J. L. *J. Appl. Phys.* **2000**, *87*, 824.
56. Sweeney, A. E.; Seal, S., *Nanoscale Science and Engineering Education: Issues, Trends and Future Directions*. American Scientific Publishers: Stevenson Ranch, CA, 2007 (in press).

Chapter 11

Integration of nanoscience into the undergraduate curriculum via simple experiments based on electrospun polymer nanofibers

Nicholas J. Pinto

**Department of Physics and Electronics
University of Puerto Rico-Humacao
Humacao, PR 00791**

Nanoscience and nanotechnology is introduced into the undergraduate curriculum via the use of electrospun polymer nanofibers. The Intermediate Laboratory core course has been modified to include several experiments on the fabrication and testing of devices and sensors made from electrospun polymer nanofibers. Students learn to use the Scanning Electron Microscope (SEM) and the Atomic Force Microscope (AFM) to measure the fiber diameter and compare it to that of a human hair. They learn how to prepare isolated polymer nanofibers and make the electrical connections for device characterization, a procedure that requires a significant amount of patience. Students also see how the extremely high surface to volume ratio of polymer nanofibers makes them very sensitive to atmospheric gases and can be used in the fabrication of supersensitive sensors. Overall, the students graduate having had real hands on experience with nanoscience and nanotechnology and are trained in the use of sophisticated instruments like the SEM and the AFM.

Introduction

The Department of Physics and Electronics at the University of Puerto Rico-Humacao (UPRH) offers a Bachelor's degree in Physics Applied to Electronics. The program is structured such that students take core courses in Physics and in Electronics with the more specialized courses (like Quantum Mechanics or Advanced Electronics Design) given as electives. Students graduate having a strong experimental background in electronic instrumentation as several core courses have a laboratory section to accompany them. The central idea in this nanotechnology initiative at UPRH is to give students hands on experience in fabricating and testing devices and sensors based on polymer nanofibers. These new experiments compliment some of the laboratory exercises done with commercially available electronic components and leave students with the additional advantage of being able to fabricate and test some of the devices in the laboratory. The experiments are currently incorporated into the second semester of the Intermediate Laboratory which is a core course and so *all* students enrolled in the program are exposed to nanoscience and nanotechnology and are trained in the use of sophisticated instruments like the Scanning Electron Microscope (SEM), the Atomic Force Microscope (AFM) and high impedance electrometers that are routinely used in this field. This paper focuses on four experiments with additional ones in various stages of development to be incorporated into the course at a later stage.

Experimental

Electrospinning is a simple technique used to prepare polymer fibers. Fibers prepared via this method typically have diameters much smaller than is possible to attain using standard mechanical fiber spinning technologies. By controlling the spinning conditions one can prepare relatively uniform fibers over several millimeters in length. Although discovered in the 1930's (*1*), electrospinning is increasingly becoming very popular in the preparation of polymer fibers either in the isolated form or in non-woven fiber mats. The basic elements of an electrospinning apparatus are shown in Figure 1. This method consists of applying a high voltage to a polymer solution placed in a hypodermic needle. As the solution is slowly pushed out through the tip of the needle via the use of a syringe pump, the force due to the large applied electric field overcomes the surface tension on the drop at the end of the needle and causes the charged solution to be ejected in the form of a fine jet. Due to bending instabilities the jet tapers to become very thin, and as the solvent evaporates, ultrafine fibers of the polymer are collected on a substrate (typically Al foil) placed a few centimeters away from the tip of the hypodermic needle. We have used this technique to prepare fibers of various polymers like polyethylene oxide (PEO), polystyrene (PS) and polyaniline (PANi) in various solvents like chloroform ($CHCl_3$), tetrahydrofuran (THF), water, and acetone. The molecular weight (MW) of the polymer is important in preparing thin fibers and we have found that for polystyrene, a MW of 212,000 yielded the smallest fibers (2). The smallest fiber that we have produced using this technique was of doped polyaniline with a diameter of 4 nm, and preparation of fibers below 100 nm is relatively simple (3). In order to capture a single nanofiber for electrical characterization a substrate was quickly inserted in the path of the jet and removed. This resulted in only a few fibers scattered over the substrate some of which make contact to the pre-patterned electrodes on the substrate and can be used for electrical characterization.

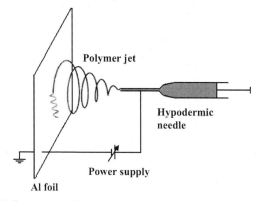

Figure 1: Schematic showing the essential components of the electrospinning apparatus.

Description of four experiments used in the Intermediate Laboratory

In the following section a brief description of these experiments is given together with figures of experimental data. Some of the experiments could be accomplished in one class period of three hours while others need more time. No more than two class periods were given to complete any experiment. Students worked in groups of two, however each student had to hand in individual lab reports containing their experimental data and the accompanying analysis. More information on the Intermediate Laboratory can be found at www.uprh.edu\~nsfnue

Fabrication of polymer nanofibers via electrospinning and their diameter comparison with that of a human hair

The aim of this experiment is to introduce students to the electrospinning technique of preparing polymer nanofibers and to use the SEM and the associated software to measure the diameter of the electrospun fiber and compare it to that of a human hair. Fibers too small to be seen with the SEM were scanned using the AFM. Most students in our program are aware that a nanometer is 10^{-9} meter but they are unable relate this size to that of common objects. Hence the focus in this experiment is to give the students direct experience in the measurement of small objects that possess "nano" dimensions. PEO or PS are two commercially available polymers that can be dissolved in many common organic solvents (viz. $CHCl_3$ or THF) and electrospun at a voltage ranging from 10 kV to 20 kV. The fibers thus formed can be collected on an Al foil for further analysis. A small piece of the Al foil with the fibers was then mounted into the SEM together with a small piece of the students' hair taken from the head. Figure 2 shows the SEM image of the polymer fibers and the hair. Students are asked to use the SEM to obtain the best image possible and take two images at low and high magnification of the hair and the fiber.

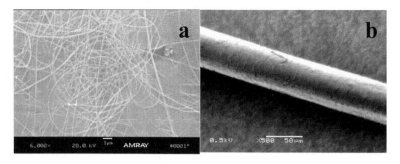

Figure 2: SEM images of (a) electrospun polystyrene (MW 212,000) nanofibers from THF and (b) human hair. The average diameter of the polymer fiber was 60 nm while that of the hair was ~40 μm.

Once the images are saved they then use the program "analySIS" to determine the diameters of the fiber and the hair. Typical fiber diameters range from several tens of nanometers to a micrometer while that of a human hair falls in the range 40 μm to 100 μm. The above experiment is very simple to perform, it can be completed in three hours and is ideally suited to introduce students to nanotechnology.

Electrical characterization of an isolated polyaniline nanofiber

Having had some experience with electrospinning from the previous experiment, students move to the next level of complexity. In this experiment they electrically characterize isolated conducting polyaniline nanofibers. The solution used for the preparation of these nanofibers was prepared as follows: 100 mg of emeraldine base PANi was doped with 129 mg of camphorsulfonic acid (HCSA) and dissolved in 10 mL $CHCl_3$ for a period of 4 hours. The resulting deep green solution was filtered and 10 mg of polyethylene oxide (PEO) having a molecular weight of 900,000 was added to the solution and stirred for an additional 2 hours. PEO was added to assist in fiber formation by acting as a plasticizer. The solution was then filtered using a 0.45 μm PTFE syringe filter. All the chemicals used can be purchased commercially. Using the electrospinning technique with an applied voltage of 8 kV, polyaniline nanofibers were captured by passing the substrate momentarily in the path of the jet. The inset to Figure 3 shows an SEM image of a single fiber bridging two electrodes..

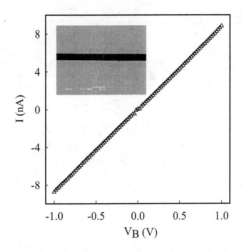

Figure 3: Current-voltage characteristics of an isolated doped polyaniline nanofiber at 300K in a vacuum of ~10^{-2} Torr. The fiber resistance was 114 MΩ and the conductivity was ~10^{-2} S/cm. Inset: SEM image of an isolated electrospun fiber making contacts to two gold leads. The lead separation was 2 µm and the average fiber diameter was 150 nm.

The length (l) and the diameter (d) of the fiber was measured using the SEM software. A Keithley Electrometer model 6517A was used to measure the two terminal current-voltage characteristics from which the fiber resistance (R) was calculated. Figure 3 shows the I-V characterisitics of a single nanofiber at 300 K and in a vacuum of 10^{-2} Torr. It was important to have the sample in vacuum to avoid the influence of air currents that lead to noisy data due to the sensitive nature of the fiber to atmospheric gases, especially the humidity. In this current experiment the length and the diameter of the fiber was 2 µm and 150 nm respectively and the measured resistance was 114 MΩ. The conductivity (σ) of the fiber was then calculated using the standard formula ($\sigma = 4l / \pi d^2 R$) which the students need to derive from elementary Physics principles and gave a value of ~10^{-2} S/cm. This is an experiment that can be completed in three hours. At the end of this experiment the students were able to isolate single polyaniline nanofibers, use the SEM for diameter calculation, make connections to electrically characterize them and compare its conductivity to that of a thin film of the same polymer

Electrospun polyaniline nanofiber gas sensors

Another experiment based on polyaniline nanofibers is the fabrication and testing of these fibers as gas sensors (*4-5*). The central idea is to demonstrate the use of these fibers as gas sensors and also that due to the high surface to volume ratio of the fibers they have a larger response and a faster response time than sensors made from thin cast films of the same material. The gas used for this purpose was methanol prepared by bubbling a constant flow of dry nitrogen gas into methanol and then having the saturated alcohol vapor flow over the fiber while the resistance was monitored as a function of time. The fabrication of these sensors is identical to that used in the electrical characterization of single polyaniline nanofibers mentioned above with the difference that the fiber is electrically characterized not in a vacuum but in a gas chamber that has been adapted for such measurements. Once the fiber has been electrically contacted and inserted into the chamber, a constant voltage is applied to the fiber and the current monitored as a function of time. First, the control dry N_2 gas is flowed over the sensor. N_2 is inert and does not react with the fiber. Once the current stabilizes then the gas is switched to methanol and when the current stabilizes again the gas is then switched to N_2 and back again to methanol in periods of 200 s respectively. Figure 4 shows a plot of the normalized resistance of the fiber as a function of time where R_{N_2} is the resistance of the fiber in the presence of N_2 prior to the first exposure to methanol. A thin film of the same polymer is also prepared on pre-patterned substrates and its resistance measured in an identical manner to that of the fiber and is plotted in Figure 4. This experiment has not been tested as yet in class but will be part of the lab next year. It is anticipated that students will complete the experiment in three hours. Students will then use the data to measure the response times and the relative response amplitude of the two samples. Typically, the fiber response is larger and also is a bit faster than the cast film. Students will also learn the reasons for the change in the resistance in the presence of methanol and the control gas which is associated extended chain conformation and with moisture adsorption. Testing sensors is important and students will realize the importance of taking the necessary precautions when measuring small currents.

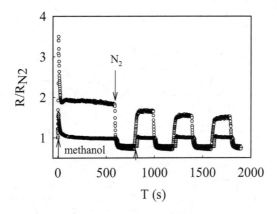

Figure 4: Normalized resistance of individual electrospun HCSA doped polyaniline nanofibers (○) and cast film of the same polymer (Δ) to vapors of methanol. The response of the nanofibers is larger and is slightly faster than the cast film due to the larger surface to volume ratio of the fiber over the film. The arrows indicate the moment when the specified gas was introduced into the chamber.

Fabrication and electrical characterization of a Schottky nanodiode

A more advanced experiment is the fabrication and electrical characterization of a Schottky nanodiode. While the fabrication of the device is the easy part, the electrical characterization and data analysis is relatively complex. In general, a Schottky diode is formed when a junction of a p-doped polymer with an n-doped inorganic semiconductor is formed. This construction can be achieved via electrochemical polymerization (*6-9*) or spin coating (*10*) the polymer onto the n-doped semiconducting substrate. In this experiment doped polyaniline nanofibers are used in the fabrication of Schottky nanodiodes. The Schottky nanodiode is prepared using an n-doped Si wafer (<111>, 1-10 Ω-m) with a 200 nm thermally grown oxide layer which can be purchased commercially and polyaniline (PANi). After prepatterning gold electrodes over the oxide via standard lithography and lift-off techniques the substrate is cleaved through the electrodes. The exposed cleaved surface has the edge of the gold electrode separated from the doped Si by the insulating oxide layer. Using electrospinning individual, charged, dry and flexible PANi nanofibers were deposited over the wafer edge making contacts to the gold and the doped Si and that are stable with no degradation or oxidation. The resulting Schottky nanodiode is formed along the vertical edge of the substrate at the nanofiber-doped Si interface (11). Figure 5(a) shows a schematic of the external electrical connections and Figure 5(b) shows a SEM image of a nanofiber at the wafer edge. The device is placed in vacuum and the current voltage characteristics are recorded with a Keithley Electrometer. Figure 6 shows the device characteristics at 300 K in vacuum. In order to quantitatively analyze the diode characteristics we assume the standard thermionic emission model of a Schottky junction as follows (12)

$$J = J_s \left[\exp\left(\frac{qV_B}{nk_BT} \right) - 1 \right] \quad (1)$$

$$J_s = A^* T^2 \exp\left(-\frac{q\phi_B}{k_BT} \right) \quad (2)$$

where J is the current density, J_s is the saturation current density, q is the electron charge, k_B is the Boltzmann constant, T is the absolute temperature, ϕ_B is the barrier height and n is the ideality factor which takes into account corrections to the original simple model e.g. image-force barrier lowering. The Richardson's constant A^* is calculated to be 120 A/K^2-cm^2. Students are taught

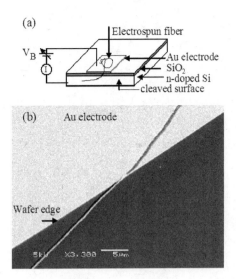

Figure 5: (a) Schematic of the Schottky nanodiode with the external electrical connections (b) SEM image of the fiber at the wafer edge. The fiber is flexible and does not fracture as it bends over the wafer edge.

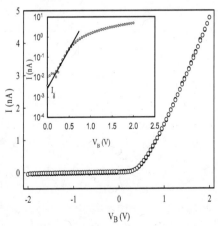

Figure 6: Current-voltage characteristics at 300 K of the Schottky diode shown in Figure 5. Inset: Semi-log plot of the forward bias current as a function of forward bias voltage used for extracting device parameters.

the theory about Schottky barriers and about band bending at the junction. The inset in Figure 6 shows a representative semi logarithmic plot of the diode current versus applied voltage under forward bias conditions at 300 K. At low biases a linear variation of the current is observed consistent with Equation (1) while the deviation from linearity at higher bias voltages generally is related to Ohmic losses due to the diode series resistance. Extrapolating the linear portion of the semi-log plot to zero bias yields a saturation current density of 5.95 × 10^{-2} A/cm^2 and the diode ideality factor calculated from the slope of the linear portion of the plot as follows:

$$n = \frac{q}{kT}\left(\frac{\partial V_B}{\partial \ln J}\right) \quad (3)$$

is $n \sim 5$. Using these equations a barrier height of 0.49 eV was calculated. Students find the data analysis the hardest part and so this experiment is allowed two class periods to complete. This is a good experiment that combines fabrication of nanofibers, devices, electrical characterization and data analysis and gives the student an overall review of polymer science, nanoscience and electronics.

Conclusions

Electrospinning is an easy method of preparing polymer nanofibers and is routinely used in the undergraduate Intermediate Laboratory at UPRH to introduce students to nanoscience and nanotechnology. Students learn to fabricate devices and sensors based on these electrospun polymer nanofibers. By comparing the diameter of polymer nanofibers to that of a human hair the students have a greater understanding of the nanoregime. Students learn to electrically characterize isolated conducting polymer nanofibers and are aware of the low power requirements in device operation and how to make careful measurements when current levels are small. Finally, students see how the extremely high surface to volume ratio of polymer nanofibers makes them very sensitive to atmospheric gases and how they can be used in the fabrication of supersensitive sensors. Overall the students graduate having had real hands on experience with nanoscience and nanotechnology and are trained in the use of sophisticated instruments like the SEM and the AFM.

Acknowledgements

This work was supported by NSF under grants EEC-NUE 0407137 and DMR-RUI 0703544.

References

1. Formhals, A. US Patent 1,975,504, 1934.
2. MacDiarmid, A.G.; Jones, W.E.; Norris, I.D.; Gao, J.; Johnson, A.T.; Pinto, N.J.; Hone, J.; Han, B.; Ho, F.K.; Okuzaki, H.; Llaguno, M. *Synth. Metals* **2001**, *119*, 27-30.
3. Zhou, Y.X.; Freitag, M.; Hone, J.; Staii, C.; Johnson, A.T.; Pinto, N.J.; MacDiarmid, A.G. *Appl. Phys. Lett.* **2003**, *83*, 3800-3802.
4. Pinto, N.J.; Ramos, I; Rojas, R; Wang, P-C.; Johnson Jr., A.T. *Sens. Actuators B* **2008**, *129*, 621-627.
5. Virji, S; Huang, J.; Kaner, R.B.; Weiller, B.H. *Nano Lett.* **2004**, *4*, 491-496.
6. Inganäs, O.; Skotheim, T.; Lundström, I. *J. Appl. Phys.* **1983**, *54*, 3636.
7. Sailor, M.J.; Klavetter, F.L.; Grubbs, R.H.; Lewis, N.S. *Nature* **1990**, *346*, 155.
8. Lonergan, M.C. *Science* **1997**, *278*, 2103.
9. Tuyen Le., T.T.; Kamloth, K.P.; Liess, H.-D *Thin Solid Films* **1997**, *292*, 293.
10. Halliday, D.P.; Gray, J.W.; Adams, P.N.; Monkman, A.P. *Synth. Metals* **1999**, *102*, 877-878.
11. Pérez, R.; Pinto, N.J.; Johnson Jr., A.T. *Synth. Metals* **2007**, *157*, 231-234.
12. Horowitz, G. *Adv. Mat.,* **1990**, *2*, 287-291.

Chapter 12

Exploring the Scanning Probe: A Simple Hands-on Experiment Simulating the Operation and Characteristics of the Atomic Force Microscope

Anthony Layson[1*], Ryan Leib[2,3] and Dale Teeters[2]

[1]Department of Chemistry and Biochemistry, Denison University, Granville, OH 43023; *laysona@denison.edu
[2]Department of Chemistry and Biochemistry, The University of Tulsa, Tulsa, OK, 74104
[3]Current Address Department of Chemistry, Berkeley, Berkeley, CA

A simple yet effective model atomic force microscope is presented for the purpose of introducing students to the field of scanning probe microscopy. The model is used to demonstrate the basic characteristics of the atomic force microscope by realistically simulating the operation of the actual instrument. Students acquire data by first using the model to scan small objects, then use a spreadsheet program to generate one, two and three-dimensional images based on the scan data. Comparisons between these images and the physical objects allow students to immediately observe how various instrumental parameters, or changes in parameters, such as sampling number or tip size, affect the resultant images. Differences in the observed images can then be used to promote discussion on other important topics such as image calibration, lateral vs. vertical resolution and tip effects.

Introduction

The term nanotechnology is used to describe a rapidly advancing and wide array of science focused on the study of nanoscale size materials and their properties. Due to the recent explosion of work in this field, there is much interest in bringing the subject of nanotechnology to the classroom. One attractive way to educate others on the topic of nanotechnology is to demonstrate the techniques that are actually used in the laboratory to study these materials. Arguably the technique that has advanced the field of nanotechnology more than any other is the scanning probe microscope (SPM) (*1*).

Several papers have focused on the use of scanning probe techniques for educational purposes (*2-9*). However, those works primarily focused on the use of the microscope as a means for analysis rather than focusing on the unique aspects of the technique itself. Furthermore, the experiments required access to actual scanning probe instrumentation, which are still not widely available at many smaller colleges and universities. Even so, the topic of scanning probe microscopy is increasingly being introduced into the classroom lecture. Yet the lack of instrument availability should not preclude this important technique from being examined in a laboratory setting as well.

There are two primary types of scanning probe instrument, the scanning tunneling microscope (STM) and the atomic force microscope (AFM). Surface examination by STM relies on electrical conductivity between the surface and the probe therefore only electrically conductive samples can be examined. The AFM relies on the force interaction between the probe and surface allowing a much wider variety of samples to be studied. We have chosen to examine the AFM in more detail due to the broad nature of the technique.

In this experiment a simple inexpensive model is constructed and used to simulate the AFM. Using data collected with the model and a spreadsheet program such as Microsoft Excel™, AFM images can be generated and analyzed in a manner surprisingly similar to that of the real instrument. With this model students can examine the various properties and characteristics of the AFM and obtain a working knowledge of the technique, which can then be applied to the study of such topics as surface science and nanotechnology. The model is designed to be very hands-on and encourages critical thinking about fundamental aspects of AFM. It is therefore applicable for upper level chemistry courses such as instrumental analysis. The concepts are easily adaptable to a more general undergraduate laboratory and can also be used as an effective classroom demonstration.

Design of the AFM Model

Figure 1a shows the setup for the model AFM. A ring stand and several tube clamps (or similar) are used to assemble the model. The cantilever is constructed from two wooden craft sticks, which have been glued to form the apex of the cantilever. A pocket laser or small flashlight shines down on a small mirror mounted at the end of the cantilever. The beam is reflected onto a wall or a screen. Care should be taken to avoid eye exposure when using the pen laser. Attached beneath the cantilever is the probe tip (inset in Figure 1b). For this model a common electrical wire nut was used. The tip is attached using double stick tape. The use of a non-permanent adhesive allows for tips to be interchanged, which can be used to illustrate how images might be affected when tips of different size and shape (aspect ratio) are used. One beneficial aspect of this model's design is that the cantilever can be either free to rotate on its pivot, mimicking the contact scanning mode, or locked in place. By locking the cantilever the magnetic analogue of AFM, magnetic force microscopy, can also be examined by attaching a small magnet in place of the tip (*10*).

The model AFM is of course much larger than an actual AFM. The cantilever length is on the order of centimeters rather than micrometers and the objects to be imaged are at the millimeter scale as compared to nano or micrometers with a real AFM. The mechanism by which data is collected however, is essentially the same. A laser beam is deflected off the cantilever onto some detection device. In the AFM photodiodes are used to measure the deflection. In this experiment the deflection of the beam spot will be monitored with a piece of paper mounted on the wall. Lined notebook paper is a good choice for the detector with the lines acting as a convenient measurement scale. Beneath the tip, a piece of paper with a simple grid is used to aid the scanning process. The grid spacing is arbitrary, and as will be discussed, is one factor in determining the image resolution during scanning. A grid size should be chosen that will enable the student to collect sufficient data points to reconstruct a good image but not so many that data collection becomes too time consuming. A grid size between 2 and 5 mm was found to work well.

The size of this model allows for the scanning of many different kinds of small ordinary objects. To aid in scanning, the objects are glued to small squares of stiff paper or cardboard. Using the grid as a guide, the sample square is scanned (parallel to the cantilever) under the tip. At each grid point the deflection of the beam on the wall is recorded. The beam deflection is recorded point-by-point for each grid space until the trailing edge of the sample square is reached. The sample is then moved up one grid space (perpendicular to cantilever) and the scanning procedure is repeated. This process is repeated until the entire sample is scanned. The Microsoft Excel™ spreadsheet program is used to produce the scanning probe image from the deflection data. A more comprehensive procedure for the cantilever construction, scanning procedure and spreadsheet data handling, can be obtained by contacting the corresponding author.

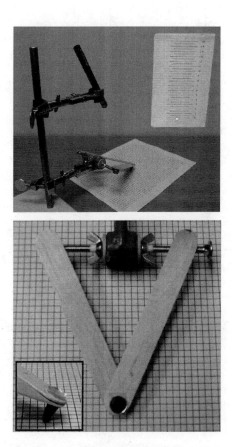

Figure 1. (a) The experimental setup for the model AFM. (b) Close-up view of the cantilever assembly. The inset in (b) shows a side view of the cantilever tip.

Pedagogical Concepts Gained from the Operation of the Model

In scanning probe microscopy, images are shown in predominantly three ways: one-dimensional line scans (cross section analysis) and two and three-dimensional topographical plots. The Excel™ program provide a means to generate one, two and three-dimensional images using data obtained from the model in much the same way that images are generated from an actual AFM. To demonstrate, the chemical symbol for potassium (Figure 2a) was made with cut pieces of a soda straw, glued to a rigid sample card. Figure 2b shows the two-dimensional (2D) view of the sample, while Figure 2c shows two different perspectives of the three-dimensional (3D) image. The topographical height scale is the same for both 2D and 3D images and is indicated by the gray scale intensity gradient in the scale bar. One-dimensional line scans are shown in Figure 2d. The open and closed arrows in Figure 2b indicate the vertical and horizontal 1D-scans, respectively.

The images of Figure 2 show that Excel™ does a good job of producing an image of the object. The 3D view is particularly useful in that it allows the student to compare and contrast the features of the generated image to the real object. The 3D image can also be rotated to allow visualization from differing perspectives. This visualization is important in understanding how images obtained in scanning microscopies differ from those in the more traditional light microscopies. The shape of the straws serves as a good example. The soda straws are smooth and round. However, the images show the straws to be almost triangular in shape with sharp sloping edges. As a result the straws in the image appear to be wider and the intersections are less defined.

The differences between the object and image, illustrate a very important principle when interpreting SPM images - this type of microscope does not provide a *real or true image* of the surface. Rather, the image is a *representation* of the surface governed by the interaction of the probe tip with the surface. As a result, the topography observed in the image can and often does differ from true surface topography. Topographical differences or distortions like those observed in Figure 2 are not restricted to this model, but regularly appear in real AFM images as well. Again, this is due to the interactions between the tip and surface and the conditions that influence those interactions. For example, the type of tip (long, thin, short, stubby), cantilever (high or low spring constant) and even the type of surface (smooth, rough, slick or sticky) will influence the tip-surface interaction. Changing scan parameters such as applied tip-force, scan direction and scan speed also play a critical role in determining an images quality, as well as how closely it represents the true surface. It is because of this that great care and some degree of knowledge about the surface being examined, is necessary when interpreting scanning probe images.

The benefit of this model is that it allows various scan parameters to be altered thereby allowing the student to see the direct affect that change has on the image. One example shown here demonstrates how altering the grid size changes image resolution. In AFM this is equivalent to changing the number of points per scan line, which in turn changes the number of scan lines per image. This is demonstrated in Figure 3 with a golf tee used as the scan object.
Figure 3b shows the tee scanned at "low-resolution" with a grid comprised of 5mm squares. A higher-resolution scan is shown in Figure 3c with a grid of 2.5mm squares. Comparison of the images show that the scan collected with more data points more closely resembles the real object. The 1D line scans more clearly illustrate this point. The low-resolution scan indicates no curvature because only two data points are sampled over the object due to the large grid size. In contrast the higher-resolution scan shows more curvature due to four points being sampled in the same area.

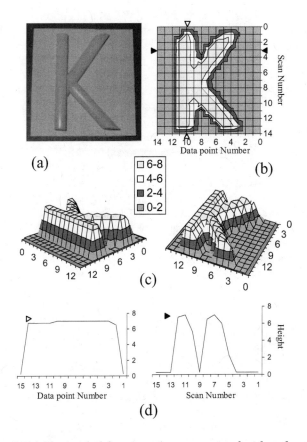

Figure 2. (a) The symbol for potassium constructed with soda straws and mounted to a stiff card. (b) 2D image generated with the Excel™ 2D surface plot. (c) 3D images generated with the Excel™ 3D surface plot. (d) Vertical and horizontal 1D line scans generated with the Excel™ line graph. The respective 1D scans are indicated in (b) by open and closed arrows. The scale bar represents the vertical height scale for all images.

This example clearly shows that the more points per scan line, the higher the resolution and the more closely the resultant image mirrors the actual surface. A comparable example of this effect is evident in the increasingly popular high definition televisions (HDTV). The image quality is superior due to an increased number of horizontal pixels and vertical scan lines. The same effect is observed in AFM. Of course, there is one caveat to increasing the resolution - speed is sacrificed. For the golf tee example, by increasing the number of scan points, it took longer to scan the sample and hence longer to complete the collection of the data necessary to generate the image. The same principle is true when using a real AFM; with more points the ultimate resolution will be higher, but scan times will most definitely be longer.

The last example that is explored here illustrates a principle of scanning probe microscopy that can be easily overlooked. Figure 4 shows an ensemble of coins mounted to a sample card. In looking at the coins it is obvious to the eye that the quarter is larger in diameter than the penny. This difference in size reflects the lateral differences (the xy-plane) in the coins dimensions. The lateral differences are easily distinguished in images obtained with the model (Figure 4b). What is not so obvious to the eye is the difference in thickness, or more appropriately, the height of the two coins. A quarter is only a fraction of a millimeter thicker than a penny. What the images in Figure 4 show is that this simple AFM is able to distinguish quite clearly the small difference in height of the two coins. This height difference illustrates the concept that vertical resolution, and the general sensitivity of the instrument in the vertical dimension, is greater than that in the lateral dimensions. This is important because it is the vertical deflection of the tip that ultimately defines the shape and structure in the final image. In fact, the term *atomic* in AFM is an excellent descriptor of this principle. The AFM has a maximum lateral resolution of a few Angstroms. This is enough to see objects as small as atoms. More significantly, the vertical resolution of an AFM is in the range of *tenths* of Angstroms, or fraction of an atom in height! This capability not only allows atoms to be observed, but allows the atoms to be distinguished from each other.

The demonstration of vertical resolution provides the opportunity to discuss in more detail some of the instrumental components in an AFM: the piezo-electric scanner and the materials from which they are made, the detectors and the detection scheme used in AFM, and the electronics needed to manipulate and control these sensitive components. With the model AFM, the component that most easily lends itself to discussion is the detector. The detection scheme used in the model AFM is quite similar to that used in a real AFM, where extremely small deflections at the cantilever (due to the tip-surface interaction) translate into much larger and easily measured deflections at the detector. While sensitive electronics are essential in accurately measuring minute deflections at the tip, simple geometry also plays a role. The vertical deflection of the laser beam at the detector becomes much larger as the detector is moved farther from the cantilever. With the model AFM the cantilever physically deflects only a few millimeters as it is scanned over the coins, but the observed laser deflection may be several centimeters or more at the paper detector. Increasing the distance between the model and the detector further increases the laser deflection, resulting in more precise measurements of the sample and ultimately higher vertical resolution. These same geometric principles also apply to the real AFM and help to explain how cantilever deflections resulting from nano to subnanometer sized surface features are readily distinguishable at the detector.

By using the model AFM, it is easy to see that lateral resolution in practicality can be improved only so far (by decreasing the grid size) before it becomes difficult or too time consuming to collect data. On the other hand, changing the measurement scale on the detector (mm spacing as opposed to inches) or changing the geometry of the instrument itself, offers an easy way to observe an increase in vertical resolution.

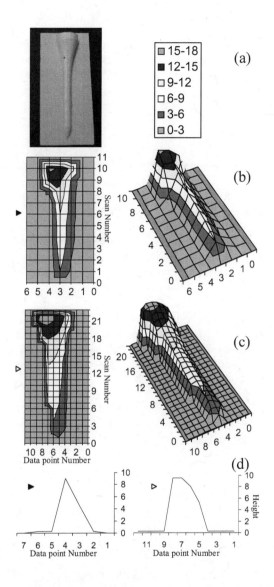

Figure 3. Shows the effect of sampling on resolution. A golf tee (a) was used as the sample object. (b) Shows low-resolution images for a scan obtained using a 5mm grid. (c) Shows the same object imaged with higher resolution using a 2.5mm grid. (d) Line scans for the low and high-resolution images are indicated by arrows in their respective 2D plots.

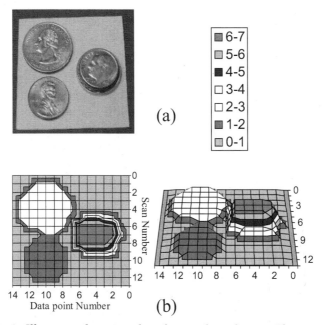

Figure 4. Illustrates the principles of vertical resolution. The quarter and penny to the left in (a) differ in height by a fraction of a millimeter. The Excel™ images clearly show the model's ability to distinguish this small difference.

Conclusions

The focus of this exercise is to introduce students to the subject of nanotechnology by demonstrating one of the cutting edge tools used in the field, the atomic force microscope. Building a scale model that closely represents the real instrument in both form and function, allows for a very hands-on approach to learning. By manually scanning the sample and recording data for each point, the students quickly learn the concept of how scanned data works to build an image. This point is further reinforced when the data is entered into a computer and the actual image is generated. It is at this point where the differences between scanning microscopes and the more familiar optical microscopes really becomes evident to the student. It also begins to show students how scientists are able to see and work with the smallest of objects.

From an instrumental point of view, the model AFM provides numerous opportunities to explore the characteristics and function of a real AFM. The examples in this paper explored the concept of how image resolution can be changed by varying various instrumental parameters. However, there are many other aspects to this technique that can be explored using the model. Using objects of known dimensions and comparing them precisely with those obtained in the graphical images can reinforce concepts of instrument calibration and calibration effects. Scanning various flat surfaces such as paper, glass or plastic, orthogonal to the tip will result in a sideward deflection of the beam at the detector simulating friction force microscopy (8). In the same manner tacky and smooth surfaces can be compared to demonstrate how tip interactions with the surface might alter an image. Scanning a simple object, then repeating the scan with the object rotated 90° demonstrates how changing the scan direction can change the resultant image as well. These are but a few of the many examples that can be used to demonstrate the characteristics of this important technique.

Acknowledgements

We would like to thank the Camille and Henry Dreyfus Foundation for financial support through the Camille and Henry Dreyfus Scholar/Fellow program. We would like to acknowledge the participants of the University of Tulsa Nanotechnology Workshops (also funded by the Dreyfus Foundation during the summers of 2002 & 2003) who were extremely helpful in the testing of the AFM model. Discussion and interactions with scientist from the NSF EPSCoR sponsored Oklahoma Network for Nanostructure Materials (NanoNet) were greatly appreciated.

References

1. Binnig, G.; Roher, H.; Gerber, C.; Weibel, E. *Phys. Rev. Lett.* **1982**, *49*, 57-61
2. Coury, L.A. Jr; Johnson, M. ; Murphy, T.J. *J. Chem. Educ.* **1995**, *72*, 1088-1088
3. Rapp, C.S. *J. Chem. Educ.* **1997**, *74*, 1087-1089
4. Glaunsinger, W.S.; Ramakrishna, B.L.; Garcia, A.A.; Pizziconi, V.J. *J. Chem. Educ.* **1997**, *74*, 310-311
5. Poler, J.C. *J. Chem. Educ.* **2000**, *77*, 1198-1200
6. Giancarlo, L.C.; Fang, H.; Avila, L. W.; Flynn, G.W. *J. Chem. Educ.* **2000**, *77*, 66-71
7. Skolnick, A. M.; Jughes, W.C.; Augustine, B.H. *Chem. Educator* **2000**, *5*, 8-13
8. Maye, M.M.; Lou, J.; Han, L.; Zhong, C. J. *J. Chem. Educ.* **2002**, *79*, 207-210
9. Zhong; C.-J., Han, L.; Maye, M. M.; Luo, J.; Kariuki, N. N.; Jones, Jr, W. E. *J. Chem. Educ.* **2003**, *80*, 194-197
10. Lorenz, J.K.; Olson, J.A.; Campbell, D.J.; Lisensky, G.C.; Ellis, A.B. *J. Chem. Educ.* **1997**, *74*, 1032A-1032B

Chapter 13

Benchtop Nanoscale Patterning Experiments

Yelizaveta Babayan, Viswanathan Meenakshi, and Teri W. Odom*

Department of Chemistry, Northwestern University, Evanston, IL 60208

>This chapter describes several nanopatterning experiments based on soft lithographic techniques such as replica molding, molding in capillaries, and contact printing and etching. The experiments take advantage of readily available commercial products: compact discs (CDs), glass slides, and optically transparent polymers such as polydimethylsiloxane (PDMS) and polyurethane (PU). Because these experiments are inexpensive and do not require specialized instrumentation, they can be incorporated into both advanced high school and undergraduate curricula.

Recent advances in nanoscience and nanotechnology have been a driving force for the development of new methods to create nanoscale structures and patterns. Typically, nanofabrication techniques fall into two classes: bottom-up and top-down. In bottom-up methods, larger structures are assembled from smaller building blocks (e.g. atoms and molecules), while in top-down approaches, structures are reduced from the macroscale into the nanoscale. Top-down fabrication techniques include electron-beam lithography, photolithography, and focused ion beam milling. Many laboratory experiments for undergraduate curricula have been developed based on bottom-up approaches,([1,2]) but top-down techniques have not received much attention because they typically require expensive instrumentation.

Soft lithographic techniques offer a convenient, cost-effective alternative to conventional fabrication methods because they do not require (i) specialized equipment, (ii) expensive clean-room facilities, or (iii) high operational costs. Soft lithography takes advantage of a soft, patterned elastomer, such as polydimethylsiloxane (PDMS) to replicate and transfer patterns from one surface to another (3). This inherently parallel process offers the ability to (i) pattern complex molecules, (ii) control chemical structure of surfaces, (iii) create channels for microfluidics, and (iv) pattern features over large areas (> 50 cm^2) on non-planar surfaces using ordinary lab facilities. Depending on the patterning technique, either molecular layers or three dimensional structures can be created in soft and hard materials such as polymers and metals. In general, the soft lithographic process consists of two parts: fabrication of an elastomeric transfer element (referred to as a mold or stamp) and the use of this patterned mold to create features defined by its relief structure. The mold is replicated from a 'master' with micro- or nanoscale features. Conventional masters for pattern replication are fabricated using top-down methods (4-6). While these techniques are versatile, they are expensive, time consuming and require specialized instrumentation and expertise. Because of these reasons, we use commercial compact discs (CDs) as masters with nanometer features.

This chapter will describe the fabrication of masters and the creation of PDMS molds used to generate patterns through three soft lithographic techniques (3,5,7). We will demonstrate how replica molding (RM), micromolding in capillaries (MIMIC), and microcontact printing (μCP) can create nanoscale patterns on the benchtop. The experiments outlined in this chapter are supplemented by an online, recipe-style, video manual that can be found on *NanoED Portal* of the NSF-sponsored National Center for Learning and Teaching at http://www.nanoed.org/courses/nano_experiments_menu.html.

Fabrication of Masters and Preparation of PDMS Stamps

Fabrication of Masters

We use commercially available relief structures such as compact discs (CDs), which have sub-500 nm features, as masters. Conventional CDs are made from a pre-patterned polycarbonate (PC) plastic layer covered with a thin sheet of aluminum (Al), and data on CDs is stored in a series of indentations packed in a spiral track. Each indentation in the PC layer is 110-nm deep and 690-nm wide separated by 1.2-μm areas, while the Al recording layer has the inverse features (Figure 1). These two layers, PC and Al, can be used to create two types of masters with complementary features.

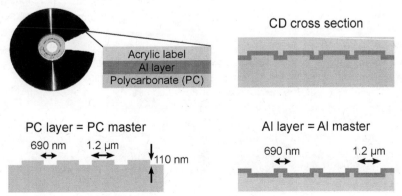

Figure 1: Schematic diagram of CD composition.
(See page 4 of color inserts.)

Materials

CD-Rs (*e.g.* Sony CD-R) and scissors.

Procedure

- Cut out a ~1×1 in² piece of CD-R using scissors.
- Peel off the Al layer from the PC layer using tweezers. The layers should separate easily from each other. Both Al and PC layers can now be used as separate masters.
- Use an atomic force microscope (AFM) to image the surfaces of the Al and PC layers (Figure 2).

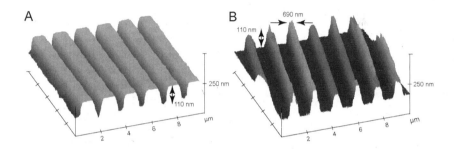

Figure 2: AFM images of (A) PC layer and (B) Al layer obtained from CD-R. Either the PC-layer or the Al-layer can be used as a master for soft lithography. (See page 4 of color inserts.)

Notes

- A piece of Scotch® Magic tape can be placed on the Al layer before peeling it from the PC layer to prevent flaking and curling of the thin Al sheet.
- An optical microscope can be used to visualize the patterns on the Al and the PC layers if an AFM is not available.

Fabrication of PDMS Molds and Stamps

An elastomeric mold, the key pattern transfer element in soft lithography, can be generated by casting a heat-curable pre-polymer against a master (in this case the PC or Al layer from a CD). The material most commonly used for fabricating molds or stamps for soft lithography is PDMS – an optically transparent, silicone-based, organic polymer. Because of its low surface free energy PDMS conforms easily to three dimensional (3D) surfaces and thus can also be used for patternng on curved surfaces. PDMS molds are formed by casting two component pre-polymer (Sylgard 184 silicon rubber base and curing agent) over the patterned master, cross-linking the elastomer by thermal curing and then peeling off the mold (Figure 3). The mold is patterned with inverse features of the master. For example, if the PC master was used to generate the PDMS mold, then the mold will have features similar to the Al master. Many elastomeric molds can be prepared from one master, and each mold can be used repeatedly for different patterning experiments.

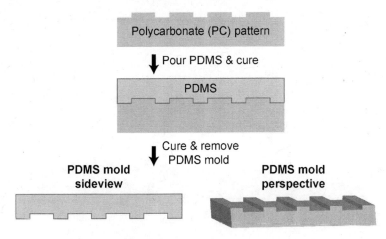

Figure 3: Scheme for making PDMS mold using PC layer. (See page 5 of color inserts.)

Materials

Sylgard 184 from Dow corning (pre-polymer and curing agent), plastic cups and forks, the Al or PC masters from a CD, scalpel, glass slides and tweezers.

Procedure

- Pour ~20 g of the Sylgard 184 pre-polymer into a plastic cup. Add 2 g of curing agent (1:10 weight ratio of pre-polymer to curing agent).
- Mix the pre-polymer mixture vigorously with a plastic fork until it is full of bubbles (1-2 min).
- Place the cup into a desiccator to degas (remove the bubbles) the PDMS for 20-40 min (or until all bubbles have disappeared).
- Pour PDMS over the PC and Al masters; be careful not to create bubbles in the polymer.
- Place the masters covered with PDMS into an oven to cure at 70 °C for 1-2 h.
- Cut out a piece of the patterned PDMS from the sample with a new scalpel.

- Remove the PDMS mold from the surface with tweezers.
- Image the mold using AFM (Figure 4).

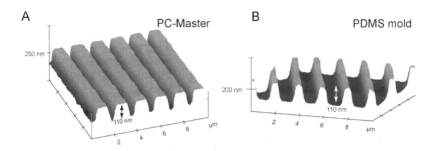

Figure 4: AFM image of (A) PC-master and (B) PDMS mold formed from (A). (See page 5 of color inserts.)

Notes

- Store PDMS molds face down on a clean glass slide.
- Use of a desiccator can be avoided by degassing the PDMS in air for a longer period of time (~2 h).
- To speed up the degassing of the PDMS in air, it can be stirred for 4-5 min instead of 2 min.
- Avoid leaving the PDMS in the desiccator for more than 2 hours because it will become very viscous and difficult to pour over the masters.
- Molds should not be thicker than 2 mm so that conformal contact between the mold and the surface can be achieved.
- PDMS should not be tacky after curing. If it is not cured completely after 1-2 h, it can be placed back in the oven until completely cured.
- PDMS molds can be cleaned by rinsing them with ethanol and drying under nitrogen or by placing a piece of Scotch® Magic tape over the patterned side and peeling the tape off the surface (use ethanol to rinse the mold after using the tape).

Soft Lithography Experiments

Replica Molding (RM)

Replica molding can produce multiple molds, replicas, and patterned surfaces from a single master in a simple and reliable way. The key step of RM is to use an elastomeric mold to replicate the relief of a 3D surface. This technique allows for highly complex structures in the master to be replicated with high fidelity down to the nm-scale. RM has been used to replicate nanostructures with features as small as 30-nm in width and 5-nm in height.(8) One disadvantage of RM is that it can only be used to create polymeric structures. Because of its parallel and inexpensive nature, RM has been used for mass production of diffraction gratings, holograms and CDs. Figure 5 illustrates the general procedure behind RM.

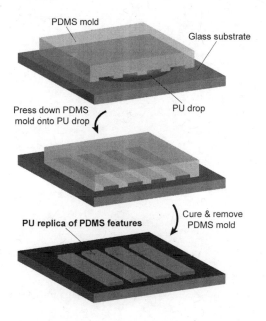

Figure 5: Scheme of RM.
(See page 6 of color inserts.)

Materials

Glass slides, PDMS molds, optically curable polymer (such as polyurethane), and UV lamp.

Procedure

- Clean several glass slides using soap and water. Rinse them with ethanol and dry under a stream of nitrogen.
- Place 2-3 drops of liquid polyurethane (PU) on a glass slide to create a puddle with an area not larger than the PDMS mold.
- Pick up the PDMS mold (either from the Al or the PC master) with tweezers and bring the patterned side of the mold in contact with the PU drop. Press down lightly on the mold to ensure that the mold makes contact with the glass.

- Place the sample ca. 5 inches under a UV lamp and let the PU cure for 10 min.
- Remove the mold with tweezers and store it on a clean glass slide.
- Look at the PU replica underneath a microscope or image with AFM (Figure 6).
-

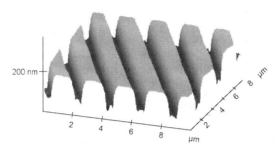

Figure 6: AFM image o the PU-replica, which is identical to the PC-master. (See page 6 of color inserts.)

Micro-Molding in Capillaries (MIMIC)

Micro-molding in capillaries (MIMIC) represents another form of RM. In MIMIC, a patterned polymeric mold is placed into conformal contact with a surface to form a network of empty channels. When a low-viscosity photocurable fluid (such as PU) is placed at the open ends of the channels, the liquid spontaneously fills the channels by capillary action. After the PU is cured under UV light, the mold is removed to leave behind a network of polymeric features. Although MIMIC is a remarkably simple technique and transfers features with high fidelity, it is limited by the viscosity of the liquid precursors that can fill the channels. Channels having dimensions smaller than 100 nm are difficult to fill by capillary action. Figure 7 illustrates the general process of MIMIC.

Materials

Glass slides, PDMS molds, optically curable polymer (such as polyurethane), razor blade, laser pointer and UV lamp.

Procedure

- Clean several glass slides using soap and water. Rinse them with ethanol and dry under a stream of nitrogen.
- Cut out a PDMS mold from the CD master. Pick up the mold with tweezers and using a laser pointer to determine the direction of the lines. Shine a laser pointer through the PDMS mold (patterned side down) above a sheet

of white paper. The diffraction pattern will be perpendicular to the direction of lines on the mold. **WARNING:** *Exercise care when working with a laser pointer. Do not place your head directly above the sample.*

- Place the mold with the patterned side facing down on a cleaned glass slide.
- Trim the ends of the mold by pressing down vertically with a razor blade on the ends that are perpendicular to the direction of the lines.
- Remove the scraps to result in open ends of the channels.
- Place a drop or two of liquid PU at one of the open ends of the channels and wait for ~10 min for the channels to fill by capillary action.
- Place the sample ca. 5 inches under a UV lamp and let the PU cure for 10 min.
- Remove the mold and store it facedown on a glass slide.
- Hold the sample with tweezers and rotate it to see the diffraction pattern, which indicates that pattern generation was successful.
- Look at the PU sample under an optical microscope or AFM (Figure 8).

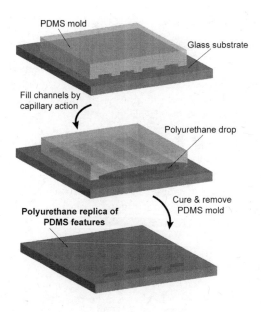

Figure 7: Scheme for MIMIC.
(See page 7 of color inserts.)

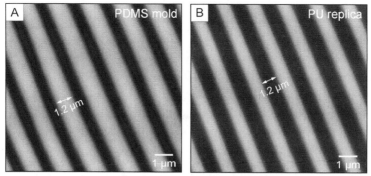

Figure 8: AFM images of (A) PDMS mold from Al-layer and (B) PU-replica of the Al-layer after MIMIC. Dark regions correspond to recessed features, and bright regions correspond to raised features.
(See page 7 of color inserts.)

Notes

- When trimming the ends of the mold, do not drag the razor along the lines, otherwise the channels will collapse and prevent the flow of PU into the channels.
- A desiccator can be used to aid the filling of the channels by vacuum.
- Caution must be practiced when using the laser pointer for determining the direction of the lines on the mold.
- The laser pointer should be held 1-2 in. above the sample, which in turn should be held 2-4 in. above a sheet of white paper.

Microcontact Printing and Etching

Micro-contact printing (μCP) is a flexible method for creating patterned self-assembled monolayers (SAMs) and for functionalizing surfaces with a variety of organic molecules. A patterned, elastomeric element is used as a stamp (the word 'stamp' is used instead of 'mold' in this section to refer to the printing functionality of the PDMS) to transfer 'ink' to metal substrates. When the PDMS stamp comes into conformal contact with a metal surface, the molecules are transferred from the stamp to the surface in the areas where the stamp is in contact with the metal. The most frequently used inks are alkanethiols, which react strongly with gold surfaces to form SAMs. These SAMs can act like masks or wet etch resists to protect the underlying metal surface. μCP can be combined with etching to produce nanostructures with the same features as the PDMS mask. One advantage of μCP over RM and MIMIC is that this technique can be used to produce not only polymeric but also metallic features. In the experiment outlined in this section, archival gold CD-Rs are used to produce complex, checkerboard metallic structures. Figure 9 illustrates μCP and etching of patterned Au CD.

Materials

Scissors, nitric acid, gold etch (Transene), plastic hemostats, archival grade Au CD-R from Diversified Systems Group, Inc.

Procedure: Preparing Au CD Sample

- Cut out a piece of Au CD-R.
- Pour ~15 mL of concentrated HNO_3 into a small beaker and place the dull side of the Au-CD facedown into the beaker. The acid will remove the protective layer from the CD. **WARNING:** *HNO_3 is highly corrosive and can cause severe burns. Handle with care.*
- Remove the CD from the beaker using plastic hemostats, rinse with water and dry with nitrogen. The protective polymer layer should peel off to reveal a patterned Au surface.

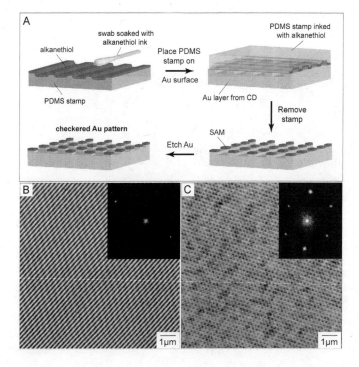

Figure 9: (A) Procedure for micro-contact printing and etching of a Au-CD. (B) Optical micrograph of unetched Au-CD and the corresponding linear diffraction pattern (inset). (C) Optical micrograph of the checkerboard pattern of the etched Au-CD and corresponding hexagonal diffraction pattern (inset). In the optical micrographs of (B) and (C), dark regions correspond to recessed features, and bright regions correspond to raised features.
(See page 8 of color inserts.)

Procedure: Micro-Contact Printing Using PDMS Stamp

- Make a 1 mM solution of octadecanethiol in ethanol, which will be the ink solution.
- Dip a cotton swab into the ink solution and rub it back and forth across the patterned side of the PDMS stamp for 10 s.
- Remove excess ink from the stamp by drying it under a stream of nitrogen for 30 s.
- Determine the direction in which the lines are patterned on the CD (you should be able to tell by the way the CD was cut; the lines on the CD run in a circular pattern).
- Bring the stamp into conformal contact with the Au surface so that the lines on the stamp are perpendicular to the lines on the Au substrate. The laser diffraction method described earlier can be used to determine which way the lines are patterned on the stamp.
- Leave the stamp on the substrate for 10 s and then remove it with tweezers.
- Breathe on the Au surface to visualize the ink transfer from the stamp to the Au sample. A diffraction pattern (Figure 9B) should be visible to indicate successful ink transfer.

Procedure: SAMs as Etch Resists

- Dilute commercial Au etchant (Transene) with deionized water (1:3 v/v) to make 15 mL of solution and stir.
- Hold the Au sample with hemostats and place it into the etching solution for 40-45 s. Remove the sample, rinse with water, and dry under a nitrogen stream.
- Place the sample under a microscope or use the laser diffraction method to visualize the Au checkerboard pattern produced (Figure 9C).

Notes

- Other alkanethiols (such as hexadecanethiol or any other long chain alkanethiol) can be used instead.
- If using octadecanethiol or any thiol that is solid at room temperature make sure it dissolves completely in ethanol. You can heat the solution but do not let the ethanol evaporate.
- Fresh thiol solution should be prepared prior to use because it produces better quality SAMs.
- The stamp should be dried after applying the ink to remove the excess thiol solution; otherwise, the excess ink will diffuse outside of the features, and no pattern will be produced.
- If the stamp does not go into conformal contact with the substrate on its own, use tweezers to press lightly on the top of the stamp.

- The angle between the stamp and the substrate can be changed during printing to produce different patterns, which can be visualized by looking at the diffraction pattern. The spacing and the symmetry of the diffraction pattern will depend on the printing angle.
- If it is hard to see the diffraction pattern, dim the lights in the room.
- If a checkerboard pattern is not obtained after etching, the experiment can be repeated by varying the time that the stamp is in contact with the Au CD (anywhere 5-30 s).

References

1. Bentley, A. K.; Farhoud, M.; Ellis, A. B.; Lisensky, G. C.; Nickel, A.-M. L.; Crone, W. C. *J. Chem. Educ.* **2005,** *82*, 765-768.
2. Winkler, L. D.; Arceo, J. F.; Hughes, W. C.; DeGraff, B. A.; Augustine, B. H. *J. Chem. Educ.* **2005,** *82*, 1700-1702.
3. Xia, Y.; Whitesides, G. M. *Angew. Chem. Int. Ed.* **1998,** *37*, 550-575.
4. Bratton, D.; Yang, D.; Dai, J.; Ober, C. K. *Polym. Adv. Technol.* **2006,** *17*(2), 94-103.
5. Gates, B. D.; Xu, Q.; Stewart, M.; Ryan, D.; Willson, C. G.; Whitesides, G. M. *Chem. Rev.* **2005,** *105*, 1171-1196.
6. Harriott, L. R.; Hull, R. *Int. Nanoscale Sci. Technol.* **2004**, 7-40.
7. Viswanathan, M.; Babayan, Y.; Odom, T. W. *J. Chem. Educ.* **2007,** *84*, 1795-1798.
8. Xia, Y.; McClelland, J. J.; Gupta, R.; Qin, D.; Zhao, X. M.; Sohn, L. L.; Celotta, R. J.; Whitesides, G. M. *Adv. Mater.* **1997,** *9*(2), 147-149.

Indexes

Author Index

Augustine, B. H., 19–48

Babayan, Y., 177–188
Bickle, J., 49–64
Bruell, C., 101–121
Burdick, J. L., 49–64

Caran, K. L., 19–48

Dadhania, M., 49–64

El-Kouedi, M., 123–133

Gerner, F., 49–64
Gorton, A., 49–64

Hoang, T. B., 49–64
Holme, T., 7–18

Iski, E. V., 123–133
Iyer, S. S., 49–64

Jadhav, S., 49–64
Jayarudu, P., 101–121
Jones, W. E., Jr., 1–3, 65–72

Knaus, K., 7–18

Larsen, R. G., 87–99
Larsen, S. C., 87–99
Layson, A., 167–176
Leib, R., 167–176

Macaluso, R. T., 75–86
Mantei, T., 49–64
Meenakshi, V., 177–188
Murphy, K., 7–18

Odom, T. W., 177–188

Pacheco, K. A. O., 1–3, 65–72
Paputsky, I., 49–64
Pienta, N. J., 87–99
Pinto, N. J., 155–166

Reisner, B. A., 19–48

Schulz, M. J., 49–64
Schwenz, R. W., 1–3, 65–72
Seth, G., 49–64
Shanov, V. N., 49–64
Smith, L., 49–64
Spatholt, A., 49–64
Sykes, E. C. H., 123–133

Teeters, D., 167–176

Watterson, A., 101–121
Willis, C. W., 65–72
Winkelmann, K., 135–154

Yeo-Heung, Y., 49–64
Yin, Y., 101–121

Zhang, X., 101–121

Subject Index

A

American Chemical Society (ACS), nanotechnology course, 2–3
Assessments
  feedback survey by University of Massachusetts, 118
  module, by University of Massachusetts Donahue Institute (UMDI), 105–108
  Nanoscience Assessment Instrument (NAI), 43–44
Atomic force microscope (AFM)
  coins, 172–173, 175f
  design of AFM model, 168–169
  effect of sampling on resolution, 171, 174f
  Excel™ program, 170–171, 172f
  experimental setup for model, 170f
  experiment simulating, 168
  golf tee, 171, 174f
  high definition televisions (HDTV), 172
  lower and higher resolutions, 171–173
  operation, 170–173
  representation of surface, 171
  three possible images, 170–171
  topographical differences, 171, 172f
  vertical resolution, 173, 175f
Atomic force microscopy (AFM)
  general science laboratory, 69
  nanotube characterization, 57
  physical chemistry at James Madison University, 32–33, 35f
  scanning probe instrument, 168
  University of Wisconsin, 70

B

Band theory
  mental effort comparison, 13f, 14f
  paired item analysis in chemistry courses, 11–14
  performance comparison on exam items in, 12f, 13f
Bimetallic iron particles
  commercial use, 101
  groundwater remediation, 102t, 103
  synthesis, 102, 111–113
Biology concepts, nanoscale science, 69–70

Books, suggestions for instructors, 16
Bottom-up engineering, topic schedule, 40*t*
Bragg's Law
  diffraction, 79
  diffraction data, 82
  Ewald sphere, 80

C

Cadmium sulfide nanoparticles
  applications, 140–141
  background, 137–139
  freshmen experiment, 137
  procedure, 139–140
  UV/vis spectrum, 139*f*
Carbon nanotubes (CNT)
  chemical vapor deposition synthesis of CNT arrays, 56–57
  cyclic voltammetry using CNT tower electrode, 58*f*
  nanotechnology topic, 55
  tower electrodes, 58–59
Chemical vapor deposition (CVD)
  carbon nanotube arrays, 56–57
  reactor, 56*f*
Chemistry classroom
  average mental effort comparison on non-nanoscience and nanoscience items, 13*f*, 14*f*
  cognitive efficiency analysis, 10–11
  nanotechnology and green chemistry inclusion, 7–8
  paired item analysis, 11–15
  performance comparison on non-nanoscience and nanoscience items, 12*f*, 13*f*
  practice exam administration, 9
  practice exam design, 9
  second-semester chemistry students, 13*f*, 14*f*
  single-semester pre-engineering students, 12*f*, 13*f*
  suggestions to instructors, 16–17
Chemistry curriculum
  introductory chemistry for science majors, 90–95
  non-majors, 95
  See also James Madison University (JMU); Undergraduate chemistry curriculum; University of Iowa
Class sizes, introducing nanoscience in undergraduate, 22
Cognitive load theory
  exam administration, 9
  exam design, 9
  mental activity, 8
Cohorts, university, 66
Coins, atomic force microscopy (AFM), 172–173, 175*f*

Color My Nanoworld
  experiment, general chemistry, 27*t*, 28
Compact discs (CDs)
  commercial relief structures, 178
  diagram of CD composition, 178*f*
  etching of patterned gold CD, 185–186
  fabrication of masters, 178–179
  mold or stamp fabrication from, 179–180
  use as masters, 178
Course curriculum, nanotechnology, 2
Critical micelle concentration (cmc)
  fluorescence emission of pyrene, 33*f*
  surfactants, 30
Current-voltage characteristics
  doped polyaniline nanofiber, 160*f*
  Schottky nanodiode, 164*f*
Curriculum, nanotechnology, 2
Cyclic voltammetry (CV), carbon nanotube tower electrode, 58*f*

D

Diffraction
  Bragg's Law, 79, 82
  crystalline materials, 76–77
  data, 82–84
  data analysis, 83–84
  electromagnetic spectrum X-rays, 78–79
  Ewald sphere, 80
  General Structure Analysis System (GSAS) software, 83–84
  generation of X-rays, 81–82
  indexing data, 83
  instrumentation, 80–81
  interplanar spacings and indices, 78*t*
  lattice centering choices, 77*t*
  nanostructures, 84–85
  powder, analysis, 84–85
  powder X-ray, geometries, 81*f*
  small-angle X-ray scattering (SAXS), 85
  tool for studying structure, 75
  unit cell dimensions, 76*t*, 77
  X-ray, pattern of NaCl and KCl, 83*f*

E

Education
  nanotechnology, 1–3
  See also Freshmen experiments
Electrospun polymer fibers
  current-voltage characteristics of doped polyaniline (PANi), 160*f*

current-voltage characteristics of Schottky nanodiode, 164f
electrical characterization of polyaniline nanofiber, 159–161
experimental, 156–157
fabrication and comparison to human hair, 158–159
fabrication as gas sensors, 161–162
schematic of electrospinning apparatus, 157f
schematic of Schottky nanodiode, 164f
Schottky nanodiode, 162–163, 165
Engineering curriculum. *See* Environmental engineering curriculum
Environmental engineering curriculum
assessment of modules, 105–108, 118, 120–121
course description, 103
module creation and description, 102–103, 111–117
module implementation, 103–104
multiple choice questions, 119–120
nanospheres for lead complexation lab, 103, 115–117
quality of modules, 106–107
student learning, 107–108, 109f
synthesis of palladized nanoscale iron particles, 102, 111–113, 119–120
trichloroethylene (TCE) degradation with palladized nanoscale Fe particles, 103, 113–115
Environmental scanning electron microscope (ESEM), nanotube characterization, 57f
Etching, microcontact printing and, 185–188
Evolutionary approach, nanoscience, 20–21, 23
Ewald sphere, Bragg's Law, 80
Excel™ program coins, atomic force microscopy (AFM), 170–171, 172f

F

Faculty, introducing nanoscience in undergraduate, 21–22
Ferrofluids
applications, 143–144
background, 141–142
pictures, 142f
procedure, 143
First year experience (FYE), university, 66
Florida Institute of Technology
general chemistry students, 135–137

See also Freshmen experiments
Fluorescence dye analysis, physical chemistry at James Madison University, 32, 33*f*
Freshman interest group (FIG), university, 66
Freshmen experiments
cadmium sulfide nanoparticles by, 137–141
ferrofluids, 141–144
gold nanoparticles and self-assembled monolayers, 144–147
magnetite nanoparticles, 141–144
nanotechnology education, 135–136
nickel nanowires, 150–151
resources, 152
silver nanoparticles, 148–150

G

Gas sensors, electrospun polyaniline nanofiber, 161–162
General chemistry
cognitive efficiency by chemistry category, 10, 11*f*
learning about nanotechnology, 135–137
See also Freshmen experiments
General chemistry lecture/laboratory
changes at James Madison University (JMU), 24–25, 28
Color My Nanoworld, 27*t*, 28
illustration of scaling, 26*f*
nanoscience laboratory activity and topics, 26*t*, 27*t*
nanoscience topics, 25*t*
General Structure Analysis System (GSAS), diffraction software, 83–84
Gold nanoparticles
applications, 147
background, 145–146
procedure, 146–147
self-assembled monolayers (SAMs) and, 144–147
UV/vis spectra, 146*f*
Gold nanorods, synthesis, 59, 60*f*
Gold surfaces, forming self-assembled monolayers (SAMs), 185–188
Golf tee, atomic force microscopy (AFM), 171, 174*f*
Graduate research, nanotechnology, 62
Green chemistry, chemistry classroom, 7–8

H

Hands-on nanotechnology

case study of quantum dots, 90–92
chemistry for non-majors, 95
content materials, 89–90
developing activities and experiments, 88
development of "nano-to-go" quantum dots, 97
first-year seminar course, 96
introductory chemistry for science majors, 90–95
laboratory experiment, 92–95
nanometer-sized cadmium selenide sulfide and cadmium selenide, 88f
quantum dot laboratory, 88–89
University of Iowa, 89
upper level undergraduate laboratory course, 96–97
High school, illustration of scaling, 26f
High school students
future outlook, 63
introducing nanotechnology, 51, 53–54
undergraduate student teaching nanotechnology, 53f

I

Indexing, diffraction data, 83
Instructors, suggestions for chemistry, 16–17
Integration, vertical, of nanotechnology, 50–51
Intermolecular forces (IMF)
mental effort comparison, 13f, 14f
paired item analysis in chemistry courses, 11–14
performance comparison on exam items in, 12f, 13f
Introduction to Nanoscale Science, learning community course, 68–69

J

James Madison University (JMU)
assessment of nanoscience experiences, 43–44
atomic force microscopy (AFM), 32–33, 35f
background of chemistry department and major, 23–24
Color My Nanoworld experiment, 27t, 28
fluorescence emission spectra of pyrene, 32, 33f
general chemistry lecture and laboratory, 24–25, 28
general chemistry nanoscience topics, 25t
illustration of scaling, 26f
laboratory development, 42–43

lessons learned from nanoscience incorporation, 41–43
literature and seminar I/II, 35–36
materials science introduction, 31
NanoManipulator tool, 33–35
nanoscience laboratories for general chemistry, 26*t*, 27*t*
new course development, 36–39, 40*t*
organic chemistry lecture and laboratory, 29–31
physical chemistry laboratory, 31–35
physics, chemistry and human experience, 36–38
scanning probe microscopy (SPM), 32–33, 34*f*
Science of the Small: An Introduction to the Nanoworld, 21, 23, 38–39, 40*t*, 44
teachers college, 43
top-down nanoscience curriculum development, 41–42

L

Laboratory development, nanoscience introduction to undergraduates, 42–43
Laboratory experiences nanotechnology course, 2
See also Freshmen experiments
Lead
complexation with carbonyl functionalized nanospheres, 105, 105*f*
engineered nanospheres for removal, 102*t*, 103
experimental complexation procedure, 115–117
multiple choice questions, 121
treatment technologies, 102
See also Environmental engineering curriculum
Learning communities (LC)
incoming freshman science majors, 67
Introduction to Nanoscale Science, 68–69
nanoscale science, 67
Supplemental Instruction (SI), 67
universities, 66–69
week-by-week events, 68*t*
Lewis structures
mental effort comparison, 13*f*, 14*f*
paired item analysis in chemistry courses, 11–15
performance comparison on exam items in, 12*f*, 13*f*
Literature and Seminar I/II, James Madison University (JMU), 35–36

M

Magnetite nanoparticles
  applications, 143–144
  background, 141–142
  pictures, 142f
  procedure, 143
Materials science, nanoscience introduction to, 31
Mental activity
  cognitive load theory, 8
  effort and exam, 9
  effort comparison on exam items, 13f, 14f
  practice exam format, 9f
Microcontact printing and etching, polydimethylsiloxane (PDMS) stamp, 185–188
Micro-molding in capillaries (MIMIC), replica molding (RM), 183–185
Middle school students
  building carbon buckyball, 52f
  future outlook, 63
  introducing nanotechnology, 51–53
Molecular conductance, self-assembled monolayer (SAM) chains, 124
Multi walled carbon nanotube arrays, tower, 58

N

Nanodiode. *See* Schottky nanodiode
Nanofibers. *See* Electrospun polymer nanofibers
NanoManipulator tool, physical chemistry, 32, 33–35
Nanomaterials
  advertised in products, 65
  freshmen learning, 136–137
Nanoparticles
  cadmium sulfide, 137–141
  gold, and self-assembled monolayers, 144–147
  magnetite, 141–144
  silver, 148–150
  topic schedule, 40t
Nanopatterning experiments
  AFM (atomic force microscopy) of polycarbonate (PC) and Al layer from contact disc (CD-R), 179f
  AFM of PC-master and polydimethylsiloxane (PDMS) mold, 181f
  AFM of polyurethane (PU) replica, 183f
  fabrication of masters, 178–179
  fabrication of PDMS molds and stamps, 179–181
  microcontact printing and etching, 185–188
  micro-molding in capillaries (MIMIC), 183–185
  nanoscience and nanotechnology advances, 177–178

replica molding (RM)
    experiment, 181–183
schematic of RM, 182*f*
scheme for MIMIC, 184*f*
scheme for PDMS mold
    using PC layer, 180*f*
Nanorods, synthesis of gold,
    59, 60*f*
Nanoscale science concepts,
    student introduction to,
    65–66
Nanoscience
  challenges to introduction
    in undergraduate
    curriculum, 21–22
  definition, 87
  general chemistry topics,
    25*t*
  top-down curriculum
    development, 40*t*, 41–42
Nanoscience Assessment
  Instrument (NAI), testing
  students, 43–44
Nanospheres
  lead complexation
    efficiency, 104, 105*f*
  multiple choice questions,
    121
Nanostructures, topic
  schedule, 40*t*
Nanotechnology
  carbon nanotube (CNT)
    tower electrodes, 58–59
  chemical vapor deposition
    (CVD) synthesis of
    CNT arrays, 56–57
  chemistry classroom, 7–8
  definition, 49–50, 87
  education, 1–3

flow-down training of
    teachers, 52*f*
future outlook, 63
gold nanorods synthesis,
    59, 60*f*
graduate research, 62
introduction to middle and
    high school students,
    51–54
safety, 55–56
semiconductor nanocrystal
    quantum dots (QD), 61–
    62
seminars for high school
    students, 53–54
seminars for middle school
    students, 52–53
topics, 54
undergraduate education,
    54–62
undergraduate student
    teaching, 53*f*
vertical integration plan,
    50–51
See also Environmental
    engineering curriculum;
    Freshmen students;
    Hands-on
    nanotechnology
Nanotechnology
  Undergraduate Education
  (NUE)
  funding, 54
  program, 20
Nanowires, nickel, 150–151
National Nanotechnology
  Initiative
  advances, 101
  educational goal, 1–2
  implementation, 1

National Science Foundation (NSF), funding nanotechnology course, 2–3
Nickel nanowires
 applications, 151
 background, 150–151
 freshmen course, 150
 procedure, 151

O

On-line resources, suggestions for instructors, 17
Organic chemistry lecture/laboratory, changes at James Madison University (JMU), 29–31

P

Palladium/iron nanoparticles
 dechlorination of trichloroethylene (TCE), 103, 104f
 multiple choice questions, 119–121
 synthesis, 111–113
Particle-in-a-box, physical chemistry, 32
Physical chemistry laboratory
 atomic force microscopy (AFM), 32–33, 35f
 changes to James Madison University (JMU), 31–35
 fluorescence dyes, 32, 33f
 upper level undergraduate, 96–97
Physics, Chemistry and the Human Experience, new course at James Madison University (JMU), 36–38
Polyaniline nanofibers
 current-voltage characteristics of doped, 160f
 electrical characterization, 159–161
 electrospun, gas sensors, 161–162
 See also Electrospun polymer nanofibers
Polycarbonate (PC)
 fabrication of masters, 178–179
 polydimethylsiloxane (PDMS) molds and stamps, 179–181
Polydimethylsiloxane (PDMS)
 atomic force microscopy (AFM) image, 181f
 fabrication of PDMS molds and stamps, 179–181
 microcontact printing and etching, 185–188
 micro-molding in capillaries (MIMIC), 183–185
 replica molding (RM), 181–183
 scheme for making PDMS mold using polyurethane layer, 180f
 soft lithography, 177–178

See also Nanopatterning experiments
Polymer nanofibers. *See* Electrospun polymer nanofibers
Polyurethane (PU)
  micro-molding in capillaries (MIMIC), 183–185
  replica molding, 181–183
Powder diffraction
  geometries, 81f
  nanocrystalline powders, 84–85
  patterns for NaCl and KCl, 83f
  See also Diffraction
Pre-engineering chemistry, cognitive efficiency by chemistry category, 10f
Primarily undergraduate institutions (PUIs)
  evolutionary approach to nanoscience, 20
  faculty, 21–22
  James Madison University (JMU), 23
  Project *DUNES (Developing Undergraduate Nanoscale Experiences in the Sciences)* development, 66, 70

Q

Quantum dots (QD)
  case study, 90–92
  developing laboratory, 88–89
  development of "nano-to-go", 97
  general science laboratory, 69
  laboratory experiment, 92–95
  semiconductor nanocrystal, 61–62
  See also Hands-on nanotechnology

R

Replica molding (RM)
  micro-molding in capillaries (MIMIC), 183–185
  soft lithography experiment, 181–183
Residential academic program (RAP), university, 66
Resolution
  atomic force microscope (AFM), 171–173
  coins, 175f
  golf tee, 174f
  vertical, in AFM, 173, 175f
Revolutionary approach, nanoscience, 20, 21

S

Safety, nanotechnology undergraduate education, 55–56

Scanning probe microscopy (SPM)
  general science laboratory, 69
  physical chemistry at James Madison University, 32–33, 34f
  See also Atomic force microscope (AFM)
Scanning tunneling microscopy (STM)
  general science laboratory, 69
  imaging of self-assembled monolayers (SAMs), 127–129
  laboratory exercise for SAMs, 130–132
  physical chemistry, 32
  scan measuring height changes across surface area, 131f
  scanning probe instrument, 168
  step by step procedure, 129–130
  STM image of coabsorbed SAM, 130f
  STM image of SAM of C_{10} thiol, 124f
  teaching, 125–127
  tip path over multi-component SAM, 128f
  typical layout, 126f
Scherrer's equation, diffraction data, 84
Schottky nanodiode
  current-voltage characteristics, 164f
  fabrication and electrical characterization, 162–163, 165
  schematic, 164f
Science of the Small: An Introduction to the Nanoworld
  James Madison University (JMU), 21, 23, 38–39, 44
  topic outline, 40t
Self-assembled monolayers (SAMs)
  concept of surfactant, 29
  future applications, 132
  general science laboratory, 69
  gold nanoparticles and, 144–147
  imaging of, with scanning tunneling microscopy (STM), 127–129
  micro-contact printing, 185–188
  molecular conductance through, chains, 124
  nanotechnology development, 123–124
  STM image of C_{10} thiol, 124f
  STM images of, for laboratory exercises, 130–131
  teaching STM, 125–127
Semiconductor nanocrystals, quantum dots, 61–62
Silver nanoparticles
  applications, 149–150
  background, 148–149
  procedures, 149

Small-angle X-ray scattering, nanostructures, 85
Soft lithography experiments
 fabrication, 177–178
 microcontact printing and etching, 185–188
 micro-molding in capillaries (MIMIC), 183–185
 replica molding (RM), 181–183
Spectroscopy
 mental effort comparison, 13f, 14f
 paired item analysis in chemistry courses, 11–14
 performance comparison on exam items in, 12f, 13f
Student learning, module assessment, 107–108, 109f
Supplemental Instruction (SI), learning community, 67
Surfactants, organic chemistry, 29–30

T

Teaching philosophy, introducing nanoscience in undergraduate, 22
Technology and Society, non-science majors, 95
Transmission electron microscopy (TEM), gold nanorods, 59, 60f
Trichloroethylene (TCE) dechlorination by nanoscale Pd/Fe, 103, 104f
 experimental, 113–115
 multiple choice questions, 120–121

U

Ultraviolet/visible spectroscopy
 cadmium sulfide nanoparticles, 139f
 gold colloid solutions, 146f
 gold nanorods, 59, 60f
Undergraduate chemistry curriculum
 assessment of nanoscience experiences, 43–44
 challenges to introducing nanoscience, 21–22
 evolutionary approach to nanoscience, 20–21, 23
 nanotechnology, 2
 revolutionary approach to nanoscience, 20, 21
 Science of the Small: An Introduction to the Nanoworld, 21, 23
 University of Puerto Rico–Humacao (UPRH), 156
 why nanoscience, 20–22
 See also Electrospun polymer nanofibers; Freshmen experiments; James Madison University (JMU)

Undergraduate nanotechnology education. *See* Nanotechnology
Universities, learning communities (LC), 66–69
University of Cincinnati (UC), vertical integration of nanotechnology, 50–51
University of Iowa
  case study of quantum dots, 90–92
  chemistry course offerings, 89
  chemistry for non-majors, 95
  first-year seminar course, 96
  introductory chemistry for science majors, 90–95
  laboratory experiment, 92–95
  "nano-to-go" quantum dots, 97
  upper level undergraduate laboratory courses, 96–97
  See also Hands-on nanotechnology
University of Massachusetts Donahue Institute (UMDI)
  feedback survey, 118
  module assessment, 105–108
University of Puerto Rico–Humacao (UPRH)
  nanotechnology initiative, 156
  See also Electrospun polymer nanofibers
University of Wisconsin, physical science, 70

W

Wastewater treatment facilities, course, 103

X

X-rays
  diffraction powder patterns, 83f
  electromagnetic spectrum, 78–79
  generation, 81–82
  powder, diffraction geometries, 81f
  small-angle, scattering, 85